How Music Works

How Music Works

The Science and Psychology of Beautiful Sounds, from Beethoven to the Beatles and Beyond

JOHN POWELL

LITTLE, BROWN AND COMPANY
NEW YORK BOSTON LONDON

For Kim

Little, Brown and Company
Hachette Book Group
237 Park Avenue, New York, NY 10017
www.hachettebookgroup.com

First U.S. edition, October 2010
Originally published in Great Britain by Particular Books, a division of the Penguin Group, August 2010

Little, Brown and Company is a division of Hachette Book Group, Inc.
The Little, Brown name and logo are trademarks of Hachette Book Group, Inc.

Library of Congress Cataloging-in-Publication Data
Powell, John.
How music works : the science and psychology of beautiful sounds, from Beethoven to the Beatles and beyond / John Powell. — 1st American ed.
p. cm.
Includes bibliographical references and index.
ISBN 978-0-316-09830-4
1. Music — Acoustics and physics. 2. Music — Psychological aspects.
I. Title.
ML3805.P69 2010
781 — dc22 2010034873

10 9 8 7 6 5 4 3 2 1

RRD-IN

Printed in the United States of America

Contents

1. So, What Is Music, Anyway?

On my first evening as a student in Birmingham I walked into the local chip shop and asked for my favorite post-pub delicacy—chips, peas and gravy. The Chinese lady looked at me quizzically and asked, "What's gravy?" I was totally flummoxed. I was used to unquestioned access to gravy back in my home town, and I had no idea how to describe the stuff... "A sort of thin brown sauce?" Fortunately the situation was saved—and a whole new world of Birmingham sophistication was opened up to me—when she smiled and said the magic words, "Curry sauce?"

This story is not about the pros and cons of gravy. The point is that sometimes we can be familiar with something we really enjoy, but have no idea what it actually is. This is the relationship most of us have with music—pleasure without understanding. To my shame, I must admit that I still don't know what goes into gravy, but I have managed to untangle some of the ingredients of music, and I hope you enjoy my explanation of how musicians manage to manipulate our moods using only string, bits of wood and lengths of tubing.

This book is not based on opinions or hopeful guesswork. It is based on real information about how musical notes are produced and what happens when they combine to form a piece of music. Many people think that music is entirely built on art, but this is not true. There are rules of logic, engineering and physics underlying the whole creative side of music. The development of music and musical instruments over the past couple of millennia has depended on a continuous interplay of art and science.

You will be glad to hear that you need no musical or scientific training to understand this book, although musicians and scientists should find plenty of things they didn't know before. The only musical skill you need is the ability to hum or sing two songs: "Baa Baa Black Sheep" and "For He's a Jolly Good Fellow"—and it doesn't matter how quietly or badly you sing them, I can't hear you. As far as math skills go, it would be useful if you can add, subtract, multiply and divide, but even these skills are not essential. Also, because I am assuming that you have no training in the subject, I will explain the meaning of any specialist words I use as I go along. This may be a little over-explanatory for the musicians and scientists among you, but I would rather be irritating to some than baffling to others.

Throughout the book I have occasionally provided details of pieces of music that might be useful to illustrate various points. Most of these examples can be watched on YouTube or listened to through other media—but they are not a necessary part of reading the book. I put them in because you might enjoy them, and because this is an excellent opportunity to advertise some of my favorite music. If you think I have explained something badly or you need more detail, please contact me by email on howmusicworks@yahoo.co.uk and I'll see if I can come up with an answer. (This address can also be used by wealthy music firms who want to bribe me with vast sums to include references to particular bits of music in future editions of this book.)

Music covers an enormous range of subjects, from the love lives of the great composers to how to build a guitar or play the trumpet. You could say that books about the history of music cover the "when" questions and most other music-related books address the "how to" problems. This book, on the other hand, deals with some of the "what" and "why" questions about music like: what is happening to the air between the instrument and your ears? And why do such things affect your mood?

Read on, and you will discover the answers to these, and lots of other questions, including:

- What's the difference between a note and a noise?
- What are minor keys, and why do they sound sad?
- Why do ten violins sound only twice as loud as one?
- Why do clarinets sound different from flutes?
- Why are Western instruments all tuned to the same notes—and why those notes, not others?
- What is harmony and how does it work?

Some of these questions have been answered in the books you will find in the physics section of the library, filed under "Musical Acoustics." The only problem is that the technical nature of the subject has meant that these books use lots of math and complicated graphs in their explanations. Books full of graphs and math have a limited readership—and this is why the only people who seem to know anything about how music works are a few badly dressed academics (and I speak with some authority here, as a badly dressed academic myself).

When I first started to study the physics and psychology of music I thought it would be straightforward. How much can there be to understand about how saxophones and harps make different sounds or why we use scales? Then I started reading. Some of the things I had thought I understood, like loudness for example, were bizarrely complicated and much more interesting than I thought they would be. To aid my own understanding I began to condense the information into simpler explanations. Eventually I realized that most of this knowledge could be presented clearly to any reader who has no musical or technical training but simply loves music. So I started to put together the notes that eventually grew into this book.

Even some top-class musicians are not familiar with the basic underlying facts about music—they play their instruments and produce the correct notes in the right order without wondering how or why their instruments were designed to produce those particular notes and not others. It's as if the musicians are acting like waiters—they deliver the meal to us—and the food is put together by chefs (composers) from boxes of ingredients, but no one knows how or why those ingredients became available in the first place.

I think it's a shame that something as popular as music has had so much mystery attached to it. In writing this book, I haven't used any math, graphs or written music, and I have kept the style conversational. By exploring the basic facts behind what notes are, and how they can move you to dance, kiss or weep, you'll discover that many of the mysteries of music turn out to be perfectly understandable—and you'll be delighted to hear that your newly acquired understanding won't stop you from dancing, kissing or weeping.

My aim is to show you—both musicians and non-musicians—that music can be understood on a very fundamental level. This level of understanding can deepen our enjoyment of music in the same way that some knowledge of how shadows are created or how perspective works can enhance our enjoyment of a painting. Some people worry that understanding more about music will reduce the pleasure they get from it, but the reverse is true. Learning how a complicated dish is prepared makes you appreciate it even more, and doesn't change how good it tastes.

Though this is a book about *all* music, I have concentrated on Western music of all types—from Frank Sinatra, U2 and Beethoven, to nursery rhymes and film music. All these styles of music, from punk rock to opera, follow the same rules of acoustics and mood manipulation.

There are complicated, overlapping layers of musical appreciation and understanding. At first glance you might think that the performing musician knows more about the music than a listener who can't play an instrument, but this isn't necessarily true. A non-musician who is a fan of the piece being played might know a lot more about how the music should sound than a performer who is playing it for the first time. As a listener, you understand a lot about music already, but a lot of your knowledge is hidden on a subconscious level. This book will help to explain and clarify all this stuff and I hope it will give you a few "ah, so that's how it works!" moments.

But enough of the preamble—let's get on with the ambling.

What is music?

The conductor Sir Thomas Beecham was always keen on sharing his views on music, for example: "Brass bands are all very well in their place—outdoors and several miles away!" This is more than a little harsh on brass bands, but it does show that music can produce strong negative feelings as well as positive ones. When it comes to music, we all have our favorites and intense dislikes, so no definition of music can include words like "beauty" and "pleasure." All we can safely say is that music is sound which has been organized to stimulate someone—which is a bit feeble really. The "someone" might be the composer only, and the "stimulation" might mean anything from joy to tears. Thankfully it is much easier to define the individual building blocks of music: notes, rhythm, melody, harmony, loudness and so on. We will be exploring all these subjects during the course of the book, and we will begin with the most basic block of all—the musical note.

A musical note consists of four things: a loudness, a duration, a timbre and a pitch. One of these features can be summed up in just one sentence but the other three will require a whole chapter or more each. "Duration" is the easy one, so let's do it now: some notes last longer than others.

The most distinctive property of a musical note is its *pitch*, so we'll start there.

What is pitch?

Pitch distinguishes a note from a noise. I will explain this at greater length over the next couple of chapters, but for now a short introduction will help to get us started.

If you hum any tune you like, you will be choosing a duration, a loudness and a pitch for each note you produce. Subtle changes in loudness and duration during a song can carry a lot of emotional information—but as we're only going to be humming the first four notes of "Baa Baa Black Sheep" we don't need to worry *too* much about that. So, try humming notes for these four words at the same loudness with the same duration for each note. All you are choosing now is the pitch. The first two notes have the same pitch; then there is a shift up in pitch to "Black"; and then "Sheep" is the same pitch as "Black."

All musical notes involve regular, repeating vibrations in the air. When you were humming any single note just now, you were producing a regular vibration with your vocal cords that was repeating itself many times a second; when I hum "Baa" my vocal cords vibrate about a hundred times a second. When I hum the note for "Black" I have to produce a higher pitch note and I do this by increasing the number of vibrations I produce every second.

So, whether the note is produced by a vibrating string or the vibration of your vocal cords, higher pitch notes involve more vibrations per second. Every melody is made up of a string of notes of different pitches.

Naming our notes

The notes on a piano or any other instrument are named after the first seven letters of the alphabet; A, B, C, D, E, F, G. Between some of these letters there is an extra note (the black notes on a piano keyboard). For example, there is an extra note between A and B, which can be called either "A sharp" (meaning "one step higher than A") or "B flat" (meaning "one step lower than B"). This fairly daft-sounding system of naming notes has been handed down to us over the past few centuries and I will explain why in chapter 9. For the moment all you need to know is that notes are named after letters and the letter might have the word "flat" or "sharp" after it. In the illustration below, I haven't the space to write the words "flat" or "sharp," so I've used the traditional symbols "♭" for flat and "#" for sharp.

All pianos are tuned to the same pitches. If you press a piano key in Helsinki and record the pitch of that particular note, and then compare it with a piano in New York, the notes will be identical. Similarly, the notes of clarinets or saxophones are the same all over the world. You might think this is obvious, but it's not so long ago that the pitches of musical notes varied from country to country, or even from city to city. The notes everyone is using nowadays were carefully chosen—but who chose them? And why?

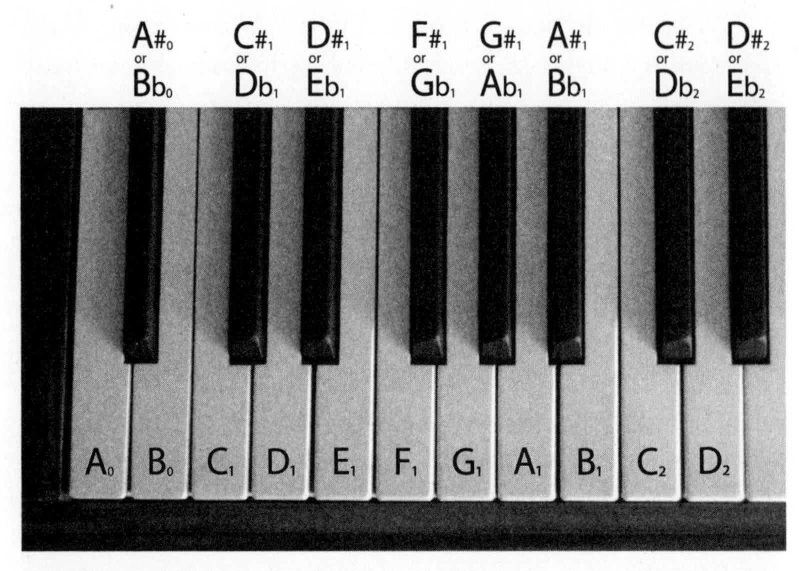

Part of a piano keyboard showing the names of the notes. As we move up the keyboard the note names repeat in a pattern which goes from A to G and starts again on every thirteenth note. There are only seven letters in the alphabet between A and G so we need extra note names ("flat" and "sharp") to cover all the notes. This peculiar method of naming notes will be explained in chapter 9. Because the names of the notes repeat, we also give them numbers. The first three notes are A_0, $B\flat_0$ and B_0. After this we have C_1 and then the number goes up each time we reach another C.

Why do we all use the same notes?

If you play a stringed instrument such as a violin or guitar, you can tighten or loosen the strings to alter their pitch. At a fairly early stage of your training you are taught how to use this tightening process to tune your instrument. This involves making the strings produce notes which are the correct distance apart in pitch. For example, the distance in pitch between any two adjacent violin strings is the same as the distance between "Baa" and "Black."

Let's say we are tuning a violin. The first step could be to tune our thickest string to the correct note, G, and then tune the other strings using the "Baa–Black" difference between each one. You would get the initial G by matching your violin note to a tuning fork* or the correct note on a (tuned) piano. But what happens if you don't have a tuning fork or piano handy?

If you are playing your instrument alone, you can choose any old note for the thickest string and then tune all the other strings to that one (making sure that the difference between any two strings is the same as the jump in pitch between "Baa" and "Black"). All you have to do to choose your first pitch is to make sure that the string is stretched tight enough to make a clear note, but not so tight that it snaps. The pitch you initially choose will not be G (unless you have perfect pitch, which I will describe in a while)—in fact, it will probably be a note between two adjacent piano keys, maybe "A and a bit" or "a bit lower than F."

As long as the difference between your strings is the same as "Baa–Black," the music you produce will sound fine and other musicians with stringed instruments could tune to match your notes and join in. However, if one of your friends is a flute player he will not be able to play along. This is because the notes on a flute (or any other wind instrument) are fixed—you can't choose any old notes on a flute. Your flute-playing friend can play, for example, an "E" or an "F" but he can't choose "E and a bit."

Let's say that you and your string-playing friends are all playing a tune that goes "E and a bit—F and a bit—C and a bit"—this will sound just as pleasant as the flute-playing E–F–C but if you both play at the same time it will sound horrible. There are only two ways to make music together:

---

* A tuning fork is a specially shaped piece of metal which produces a specific note when you hit it.

1. You violin players must hold the flute player down while one of you saws a few tenths of an inch off the end of his flute. Then you'll have to file all the holes until they are in suitable positions for this new, shorter flute; or
2. You can all tune your violins to match the flute notes. Once you have done this, any other instrument can join in with you because you are now playing the standard notes.

These standard notes are no sweeter or more musical than any other group of notes. They are only correct because someone had to decide how long flutes and other wind instruments should be. (The length of such instruments determines the pitch of the notes they produce.) In the past, things were very confused—flutes made in different countries were all slightly different lengths—which meant that a German flute player couldn't play along with an English one unless he bought an English flute. After a lot of argy-bargy about which length was the best, it was decided that a bunch of experts in badly cut suits would form a committee and decide once and for all on a group of notes that everybody would use from then on. After a lot of expert discussion (which sounds a lot like argy-bargy), they decided on the notes we use today at a meeting in London in 1939. So now, all over the world, flutes and all the other Western instruments such as violins and clarinets, guitars, pianos and xylophones have a set of standard notes.

Nowadays, if someone says "I have *perfect pitch*" they mean that the pitches of these standard notes are fixed in their long-term memory, and this peculiar ability is the subject of the next chapter.

2. What Is Perfect Pitch and Do I Have It?

Let's imagine three people are singing in the shower—no, not all in the same shower, this isn't that sort of book. These three people are all singing in their (otherwise silent) bathrooms on different floors of an apartment building.

On the second floor we have Kim Normal: she has a gin and tonic in one hand and is belting out "Dancing Queen" by Abba at the top of her untrained voice. If we taped her song and compared it to the original recording, we would discover two things:

1. Although the notes go up and down in pitch at the right places, they sometimes jump a little too far and sometimes don't jump quite far enough. This is how most of us sing (which is why we should stick to our day jobs).
2. The note she started on was not the same note that Abba started on. In fact, the note she started on doesn't appear anywhere on a piano keyboard (why should it?). It's just some note she picked from the middle of her vocal range and, if you checked, you would find that it was somewhere between two adjacent notes on a piano. Once again, this is what most of us do.

Up on the seventh floor lives James Singer: he is a trained member of his local church choir but he doesn't have perfect pitch. Fortunately for this discussion, he is also singing "Dancing Queen." If we compared his tune with the original, we would find that his vocal leaps up and down are very accurate. However, as in the case of his downstairs neighbor, the note he started on was not the same as the Abba original—it was one of those "in between" notes that most of us choose when we sing.

Up in a bathroom on the fifteenth floor, Cecilia Perfect is also reliving the 1970s, singing (wait for it)... "Dancing Queen." Cecilia is a trained singer who also happens to have perfect pitch (or *absolute pitch,* as it is also known). When we compare her rendition with the original we will find that, not only are her vocal leaps accurate, but she started on exactly the right note. This, of course, means that she is singing all the same notes as the original Abba song.

Cecilia's performance is remarkable and quite rare (only a very small percentage of people have perfect pitch), but it is not necessarily a sign that she has any special musical talent. It is possible that James is a better singer and that, if you wheeled a piano into his bathroom and played the first note of the song, he would be able to start from there and, like Cecilia, sing exactly the same notes as Abba.

What Cecilia is demonstrating is that she has memorized all the notes on a piano (or flute, or some other instrument) and it is just about certain that she managed this incredible memory feat before she was six years old. Young children remember things far more effectively than anyone else, which is how they learn to talk and acquire other skills (one minute they are sitting in the garden eating worms and going "ga ga goo goo" and a few months later they are strolling around making sarcastic comments about the quality of the cookies).

If you teach a small child a song, she will learn the tune and the words. A tune is not made up of specific notes—tunes simply involve a series of upward and downward jumps in pitch with a certain rhythm. "Baa Baa Black Sheep" sounds just as good whatever note you start on—and, don't forget, nearly all of us start on a note which is between two piano notes.

It is only when the tunes are produced on an instrument that the child might start to develop perfect pitch. If one of the parents plays the same notes on a piano each time she sings "Baa Baa Black Sheep," the child may start to remember the actual notes involved rather than just the up and down jumps of the tune. Eventually the child could build up a whole mental library of all the notes on a piano. If this happens, she might also learn that each of the memorized notes has a name such as "the F above middle C" (*middle C* is the C near the middle of a piano keyboard).

An interesting point here is that, although perfect pitch is rare in Europe and the USA, it is far more common in countries such as China and Vietnam, where the language involves an element of pitch control. The sound you make to produce a word in these tonal languages is a cross between singing and speaking. The pitch at which you "sing" a word in a language like Mandarin is vital to communication: each word has several unrelated meanings depending on its pitch. The word "ma," for example, means "mother" if you sing/say it at a high, level pitch—but it means "hemp" if you start at a middle pitch that then rises; or "horse" if you start lower then fall and rise. If you start high and let the tone fall you are saying "lazy." So an innocent question such as "Is lunch ready, Mother?" could easily become "Where's my lunch, you horse?" if you get the pitches wrong. As this sort of mistake could result in a catastrophic failure of lunch supply, young children learning these tonal languages pay much closer attention to pitch than Westerners do—and young children who concentrate on pitch a lot are more likely to acquire perfect pitch.

The reason why very few Westerners develop this note memory is because it isn't very useful to us—in fact, it can be a bit of a pain to have perfect pitch because it makes the whistling or singing of most people sound terribly out of tune. If you are an orchestral violinist, perfect pitch could be helpful in tuning your instrument to the correct pitch in the taxi on the way to a concert. If you are a professional singer, you could always be sure you were practicing the correct notes even if you were walking in the countryside—

but those are about the only benefits. This lack of usefulness is one of the reasons that musical training never involves any attempt at perfect pitch acquisition. The other main reason is that it is very difficult to achieve after the age of six.

Having said all that, quite a few musicians (and some real people) have partial perfect pitch. What I mean by this is that they have remembered one or two notes. For example, most of the musicians in an orchestra have to tune their instrument at the beginning of each concert (unlike the smug, perfect-pitch violinist who can do his alone in the taxi). They always use the note "A" for this purpose. One instrument (usually the oboe) plays an "A" and the other players adjust their instruments so that their "A" sounds the same. (This produces the dreadful wailing racket you hear just before an orchestral concert.) This repeated concentration on the note "A" can lead a few of the musicians to remember it.

Other examples of partial perfect pitch are also related to repeated exposure to a particular note or song. Sometimes nonmusicians can experience this and remember a note or notes even though they don't know the names of them. Try it for yourself. Get out one of your favorite songs and sing what you think will be the first note to be played, and keep singing it or humming it as you start your CD player. You never know, you just might have partial perfect pitch.

This partial perfect pitch is not as surprising as it might seem at first. We can all remember a note for a few seconds (try this with your CD player) and a repeated short-term memory can sometimes develop into a long-term memory.

By the way, your singing or humming will probably be much more accurate if you stick a finger in one of your ears—which is why you will see some solo singers doing this. This works because we are designed not to hear our own voices too loudly, in case they drown out any other noises we should be paying attention to—lions, avalanches, the last-call bell, etc. Sticking a finger in your ear improves the feedback between your mouth and brain and helps you monitor your own pitch much more carefully. You may have noticed that your voice–ear feedback also improves if you have blocked sinuses, which can be very annoying. (I once made the mistake of complaining about this to my girlfriend. "My voice sounds really loud and it's getting on my nerves." Her response was a single eyebrow twitch and, "Now you know what the rest of us have to put up with . . .")

I want to go back to our three singers and imagine what's going on in their heads as they sing, but first you need to know that the jumps in pitch between the notes in a tune are called *intervals* and the different intervals have names that describe how big they are. The smallest interval is the distance between two adjacent* piano keys and is called the *semitone*, twice this jump is called a *tone* (not surprisingly). You don't need to know the names of all the intervals, but you can find them—and a trick showing you how to identify them—in "Fiddly Details" at the back of the book (p. 244).

So what are our singers subconsciously thinking as they start singing the song?

Kim Normal's brain is sending out the following signals:

- sing any old note
- down a bit for the next note
- up a bit for the next note, etc.

---

* If you look back to the photo of the piano keyboard in chapter 1, you will see that the word "adjacent" is a little complicated for a piano. All the white keys look adjacent to each other because the black keys are not long enough to separate them properly. The fact that the black keys are short is merely to help with the ergonomics of the instrument. As far as the sound is concerned, the white notes B and C are adjacent to each other, but F and G, for example, are not—they are separated by F#.

James Singer's brain is sending out the following signals:

- sing any old note
- down a whole tone for the next note
- up three semitones for the next note, etc.

Cecilia Perfect's brain is sending out the following signals:

- sing C sharp
- down to B
- up to D, etc.

But, as I said earlier, the fact that Cecilia's notes agree with the ones chosen by a committee in 1939 doesn't mean that she is a better singer than James. Being a good singer is not just a matter of hitting the right notes—you have to sing them clearly, with the appropriate stress, and you have to make sure that you don't run out of breath before the final note of a phrase ends. On top of all this, the quality of your voice is affected by the shape and size of the equipment you have: your vocal cords, mouth, throat and so on. Almost any one of us could be trained to be a reasonable singer, but to be really good you need training *and* the correct equipment.

Weirdness and pedantry

In chapter 1 I mentioned that German flutes used to be a different length from English ones and this meant that German orchestras and English orchestras would be playing different notes. In fact, every country (and even some cities) had their own notes. In the nineteenth century an "A" in London would be more like an "A flat" in Milan and a "B flat" in Weimar. We know this because historians have uncovered various tuning forks from these confused times and we can also compare the notes of church organs and flutes from different places. Just to add to the chaos, the local standard notes also went up and down in pitch from decade to decade.

Imagine the scene: it is 1803, Anton Schwarz, the famous German singer, meets Luigi Streptococci, the famous Italian singer, in a pub in Bolton:

"Hey Luigi, you're singing every note flat—I know because I have perfect pitch."
"No, Anton, it's you—you're singing sharp. I know because I *truly* have perfect pitch."
"No, you're wrong."
"No, *you're* wrong."
"Flat, flat, flat."
"Sharp, sharp, sharp."

And so on—until the landlord chucks them out of the pub because neither of them is singing in tune with his piano (which is tuned to Bolton standard pitch for 1803). No wonder we used to have so many European wars in those days.

This is a weird situation. Professional musicians were (and still are) often trained from a very early age, and some of them would have developed "perfect" pitch, which agreed with the pitch chosen by a local piano tuner or organ builder. As soon as they began to travel they would discover other highly trained professionals with different "perfect" pitch. It's a bit like everyone declaring that their favorite shade of pink is the perfect pink. All these "perfect" pitches were equally valid. To have perfect pitch all you need is a set of pitches etched into your long-term memory. You don't even need to know what the notes are called—you might have stored all the notes on your mom's piano without ever being told that this one is B flat and that one is D, etc.

Nowadays, people with perfect pitch have usually memorized the standard Western pitches that were decided on in 1939 because that's how all pianos, clarinets and other Western instruments are tuned. This means that, if you have it, your perfect pitch is the same as everyone else's. Most people with perfect pitch will also know the names of the notes involved because they generally acquired their perfect pitch during some sort of musical training at an early age.

The historical facts make life difficult for the musical pedants among us. A typical pedantic view would be that we should play Mozart's music exactly as he *wrote* it. Another, equally understandable, pedantic view would be that we should play Mozart's music exactly as he *heard* it in his head as he wrote it down. Now here we have a problem, because although Mozart had "perfect" pitch, the notes he had memorized were not the same as those chosen by the committee in 1939. In fact, the note we know as "A" would be called a "slightly out of tune B flat" by Mozart (we know this because we have the tuning fork Mozart used). So when we listen to Mozart's music nowadays, we are hearing it all about a semitone higher than he would have intended—a fact which is guaranteed to annoy some musical pedants. Some of his most difficult, high-reaching songs would actually be much easier to sing if we lowered them in pitch by a semitone, which is closer to how Mozart intended them to sound. On the other hand, this would involve writing out all the music again in a lower key, which would irritate an entirely different set of pedants.

So if you are ever discussing perfect pitch, you need to bear the following points in mind:

- If people have perfect pitch, it merely means that they memorized all the notes on a particular instrument before they were around six years old. These people generally have high levels of musical skill, but this has nothing to do with their perfect pitch ability (which is rather useless). They usually have excellent musical skills simply because they started their musical training before they were six years old. Most musical skill comes as a result of training rather than inspiration: the earlier you begin, the better you will be.

• Any talk of someone having "perfect" pitch before 1939 doesn't tell us anything about the pitches of the notes involved, because there were no agreed international standard notes. On the other hand anyone who had local "perfect" pitch was probably a very good musician because he had obviously begun his musical training very early.

As to the question of whether or not you have perfect pitch, it's easy to find out by the method I mentioned earlier. Pick out a few of your favorite songs from your CD collection and try to sing the first note of each one before you play it. (Remember to put a finger in one of your ears so you can hear yourself more clearly, and don't wait for the first word to be sung because the introductory music will warn you what note is coming up. What you have to sing is the very first note on the track.)

- If you get all the notes right you have perfect pitch;
- if you get some of them right you have partial perfect pitch;
- if you think you got them right, but no one else in the room agrees, you should get some sleep and try again in the morning when you have sobered up.

3. Notes and Noises

Every day you will hear millions of sounds and only a few of them will be musical notes. Usually, musical notes are created deliberately from a musical instrument, but they can be produced in non-musical situations—when you "ping" a wineglass or ring a doorbell, for example. Whenever and however they are produced, musical notes sound different from all other noises.

What's the difference between a musical note and any other sort of noise? Everyone you know will have some sort of answer to this question, but most of them will be based on the idea that musical notes sound... er... musical and other noises are... er... not musical.

Music plays upon our emotions and can enhance or change our mood. A good example of this is the way that film sound tracks give us clues as to how to respond emotionally to the scene we are watching—romance, humor and tension are all magnified by the accompanying music. This link between music and emotion might make us think that notes themselves have an emotional content and that they are in some way mysterious and magical sounds. In fact, there really is a genuine, simple difference between musical notes and all other noises, but it has nothing to do with emotion—a computer could spot the difference every time.

Sound

If you throw a stone into a flat, calm pond you will disturb the surface of the water and create ripples which travel away from the initial splash. Similarly, if you click your fingers in a quiet room, you will disturb the air and ripples of disturbance will move away from your hand.

In the case of the stone in the pond, the ripples involve a change in the height of the water and our eyes can clearly see what's going on: the height of the water goes up–down–up–down–up–down as the ripples travel away from the splash.

When you click your fingers (or make any other sound, including a musical note), the sound ripples traveling toward your ears involve changes in the pressure of the air. We can't see these ripples but our ears can hear them. When the ripples reach our ears, the air pressure goes up–down–up–down–up–down and this makes our eardrums go in–out–in–out–in–out at the same rate—because our eardrums are like tiny, flexible trampolines which are easily pushed in and out by changes in the air pressure. Your brain then analyzes the in–out movement of your eardrums and decides what's going on—is it time to run away or time to order dessert?

Non-musical noises

As we grow up we become very skilled at identifying and interpreting noises—the boiling of a kettle, someone buttering toast or chopping wood, the playful click of the ATM as it eats your debit card—we accumulate a vast library of sounds to help us work out what's happening around us.

If we could see the pressure ripples of these non-musical sounds, we would notice that they were very complicated. The overall ripple would be created by all sorts of things happening at the same time: the sound of a door closing might involve vibrations of the door, the lock, the wall and the hinges—and each of these complicated individual vibrations would join together to produce an even more complicated set of ripples of pressure in the air.

Let's imagine that we can see the pressure ripples produced by a door closing. In the left-hand part of the illustration below I have drawn out possible ripple patterns* for the door, lock, wall and hinge separately. On the right of the illustration I have joined them all together into the overall ripple shape which pushes our eardrums in and out as the door closes.

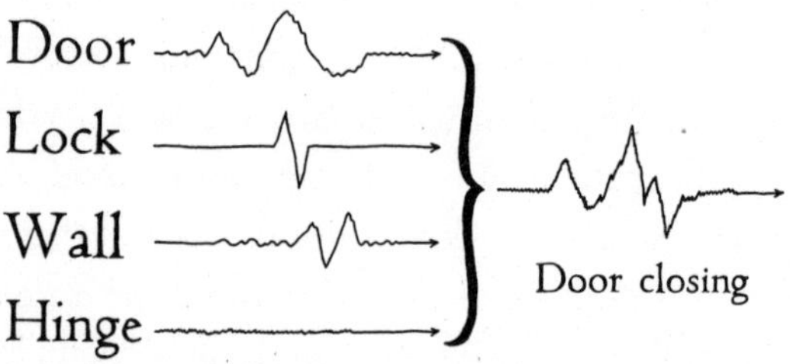

Several pressure ripples join together to make a noise. The sound ripples made by a door, lock, wall and hinge join together to make the overall noise of a door closing.

The noise ripple shape which eventually arrives at the eardrum is extremely complicated because it is made up of a chaotic group of individual ripples which have no relationship to each other. This is true of all noises which are not musical notes.

---

* We could find out what the actual ripple shape for a closing door looks like by attaching a microphone to a computer and asking the computer to draw the changes in pressure experienced by the microphone (a microphone acts rather like an ear—it has a small part inside which moves in and out as the pressure of the air goes up and down). The actual ripple pattern would probably be even more complicated than the one I have drawn here.

Musical notes

Musical notes are different from non-musical noises because every musical note is made up of a ripple pattern which repeats itself over and over again. In the illustration below there are some examples of the ripple patterns of notes produced by different instruments. To be a musical note, it doesn't really matter how complicated the individual ripples are, as long as the pattern repeats itself.

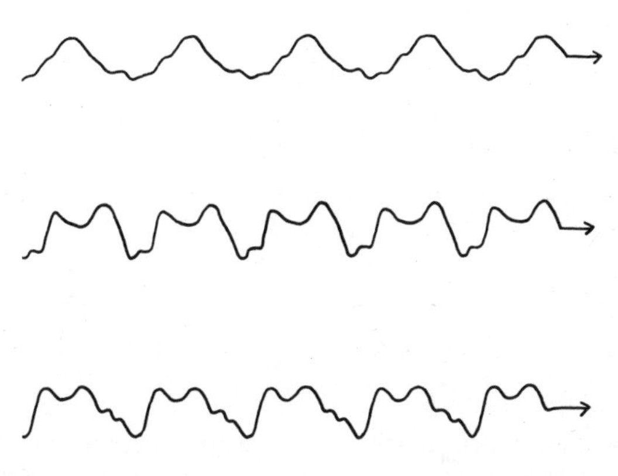

Musical notes are made up of ripple patterns which repeat themselves over and over again. These ripple patterns are actual recordings of notes produced by a flute (top), a clarinet (middle) and a guitar (bottom). (Source: Exploring Music *by C. Taylor (Taylor & Francis, 1992))*

Our eardrums flex in and out as the pressure ripples push against them. However, our eardrums can't respond properly if the ripple pattern repeats itself too quickly or too slowly—we can only hear patterns which repeat themselves more often than twenty times a second but less often than 20,000 times a second.

Musical notes don't need to be made by musical instruments, in fact, anything which vibrates or disturbs the air in a regular way between twenty and 20,000 times a second will produce a note.

High-speed motorbike engines or dentists' drills produce notes. In the song "The Facts of Life," the band Talking Heads uses what sounds like a compressed air-powered drill to produce one of the notes of the background accompaniment. This combination of music and engineering fits well with the lyrics, which compare love to a machine.

Musical instruments are simply devices which have been designed to produce notes in a controlled way. A musician uses finger movement or lung power to start something vibrating at chosen frequencies—and notes are produced.

Good vibrations

When things vibrate they generally do so in lots of different ways at the same time. For example, if you are chopping a tree down, every time the ax hits the tree, the various branches all vibrate at different rates in different directions. Each of these vibrations creates a ripple pattern in the air, but they do not join together to create an overall repeating pattern—so we hear a noise rather than a note. Musical notes are only created when all the vibrations collaborate to produce a regularly repeating ripple, like the ones in the illustration above. For this to occur, the vibrating object must only produce ripples which are strongly related to each other and can join together in an organized way. This happens most easily if the vibrating object has a very simple shape. One of the simplest

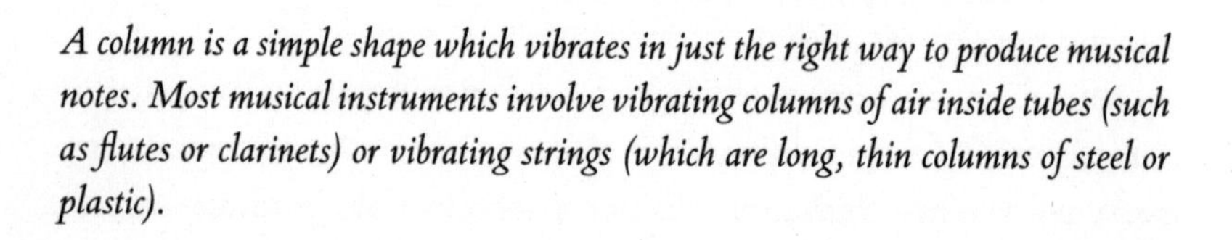

A column is a simple shape which vibrates in just the right way to produce musical notes. Most musical instruments involve vibrating columns of air inside tubes (such as flutes or clarinets) or vibrating strings (which are long, thin columns of steel or plastic).

possible shapes is a column or rod, like the one shown in the drawing. As we shall see shortly, columns vibrate in exactly the right way to produce musical notes, which is why most musical instruments involve the vibration of columns of air inside tubes (flutes, clarinets, etc.), or vibrating strings (strings are just long, thin columns made of plastic or steel).

As far as strings go, the purest, loudest notes come from new strings, when their shape is as close to a column as possible. After they have been played for a while they get bruised and damaged and become an imperfect column shape. When this happens the strings become quieter and the notes they produce can become rather vague in pitch. Professional musicians change their strings every few weeks, but when I was a student I always used to leave mine until the strings were almost untunable before I bought a new set—because you can buy a lot of chips, peas and curry sauce for the price of a set of guitar strings. Generally you have to buy a whole set, because if you just replace a couple of the worst ones you'll have two fresh new strings which sound louder and brighter than the others. And it's no good trying to cheat by buying two new strings and artificially aging them with chip fat and curry sauce—the symmetrical column shape of the new string will shine through such admirable attempts at household budgetry.

We can learn a lot about musical notes by looking at how a single string vibrates when it is plucked. When you pluck a guitar string it moves back and forth hundreds of times every second. Naturally, this movement is so fast that you cannot see it—you just see the blurred outline of the moving string. Strings vibrating in this way on their own make hardly any noise because strings are very thin and don't push much air about. But if you attach a string to a big hollow box (like a guitar body), then the vibration is amplified and the note is heard loud and clear. The vibration of the string is passed on to the wooden panels of the guitar body, which vibrate back and forth at the same rate as the string. The vibration of the wood creates more powerful ripples in the air pressure, which travel away from the guitar. When the ripples reach your eardrums they flex in and out the same number of times a second as the original string. Finally, the brain analyzes the movement of the eardrum and thinks, "That idiot next door is practicing his guitar again."

The number of times the string vibrates back and forth in one second is called the *frequency* of the note (because this is how *frequently* the string goes back and forth). One of the first people to do serious scientific research on frequencies was a bushy-bearded German called Heinrich Hertz, back in the 1880s. The scientists and musicians who worked on acoustics later found that they needed a short way of saying "the string had a vibrational frequency of 196 back and forth cycles per second." First of all they shortened it to "the string had a frequency of 196 cycles per second,"* but even bearded scientists could see that this wasn't exactly a snappy way of saying it. Eventually (in 1930) someone came up with the idea of using the name "Hertz" to describe the number of cycles per second, so now we say "this string has a frequency of 196 Hertz." Usually we shorten "Hertz" to "Hz" when we write it down: "the string had a frequency of 196Hz." (I'm sure that Dr. Hertz and his beard would both have been very pleased about being honored in this way—and I wish I could think of something which could be measured in "Powells.")

In chapter 1 I explained that the pitch (or frequency) of every note we use was decided by a committee in 1939. Although they did not specifically discuss guitars, the decisions they made applied to all instruments, so we know, for example, that the second thick-

* When we say "cycles per second," we mean how many times the string moved through a *complete* cycle in one second. A complete cycle would be, for example, starting in the middle, moving over to the right, back to the middle, over to the left and then back to the middle.

est string on a guitar—the "A" string—should vibrate back and forth 110 times a second.

So, John Williams (the classical guitarist's classical guitarist) has just twanged the "A" string on his guitar for us—and, on a properly tuned guitar, the "A" string has a *fundamental frequency* of 110Hz. A simplified view would be to say that the string is now moving to and fro 110 times every second and our eardrums are doing the same. But this is not the full story. The string is, in fact, vibrating at lots of frequencies simultaneously and 110 times per second, or 110Hz, is just the lowest of the frequencies involved. The other frequencies are multiples of this fundamental frequency, i.e., 220Hz (2 x 110), 330Hz (3 x 110), 440Hz (4 x 110), 550Hz (5 x 110), etc. When the string is twanged we hear all these frequencies at once—but the effect on our hearing system is that of an overall ripple pattern repeating itself 110 times every second.

So now we have a few very interesting questions:

1. Why does the string vibrate at lots of frequencies rather than just one?
2. Why does the string choose to vibrate only at frequencies which are related by whole numbers to the fundamental frequency (two times, three times, four times, etc.)?
3. Why do we hear one overriding fundamental frequency when all these other frequencies are joining in?

Let's look at the situation from the string's point of view. Before being twanged, the string was in a happy, stable state, i.e., in its stretched condition it occupied the shortest route between two points—a straight line. The twanging involves stretching it a bit more before letting go. This is extremely annoying for the string and it immediately tries to rush back to being a straight line as soon as we let go. Unfortunately, as it gets somewhere near its original position, it finds that it is traveling too fast to stop—so it overshoots, stretching itself in the opposite direction, then it changes direction and tries to hurry back toward its straight line shape—and continues to overshoot back and forth until it runs out of energy. The string runs out of energy because, as it moves to and fro, it has to push air out of the way all the time, and it's also passing its vibrational energy onto the wooden body of the guitar or violin.

You might think that the string would simply spring from side to side, from one smooth curve to another—as drawn on the left in the illustration below. But it cannot do that because, if it only went from one curve to another, the string would need to *begin* its journey as a smooth curve. In fact, a plucked string begins its jour-

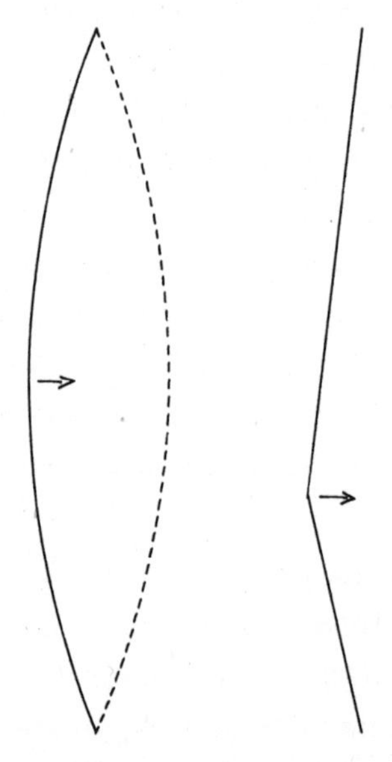

If we wanted to make a string vibrate at only its fundamental frequency, we would have to start it off in the correct shape—a smooth curve like the one in the drawing on the left. In fact, when you pluck a string, it starts off as two straight lines which meet at your finger, as in the drawing on the right.

ney as a kinked line—two straight lines which meet at your plucking finger—as in the drawing on the right of the illustration. Once you let go of it, the string won't have a chance to organize itself into a nice gentle curve—all the various bits of the string will just race off as fast as they can.

So now our string is in a bit of a quandary. It needs to move once it is released, but it cannot move as a simple curved line because it's not starting off as a simple curve. The answer to this problem is that the string starts to vibrate in several ways at the same time.

This might sound odd. How can a string do several things at once? Actually it's not that difficult. Imagine you are sitting on one of the children's swings in the park—one of those old-fashioned ones with metal chains and a wooden seat. If you swing gently backward and forward, then each link in the chain will also move backward and forward. The ones near the seat move most and the ones near the bar at the top don't move much at all, but we can say that the chain as a whole is responding to a single instruction: "Move gently back and forth." Now slap the chain in a forward direction at a height level with your ear. The chain will now be wiggling quickly backward and forward as well as swinging gently in the same direction. The chain will be following two instructions: 1. swing slowly to and fro; and 2. wiggle quickly to and fro. You can start to twist as you move, to add a third instruction, and you could add other sorts of movement if you really wanted to feel dizzy. The point is that a chain or rope or string is capable of obeying several instructions at once.

Although our obedient guitar string can follow lots of orders simultaneously, it wisely only accepts instructions which allow it to keep its ends stationary—because the ends of the strings are attached to the guitar and cannot move.

Have a look at the next illustration. Here you can see a photograph of my favorite classical guitar—and next to it are some

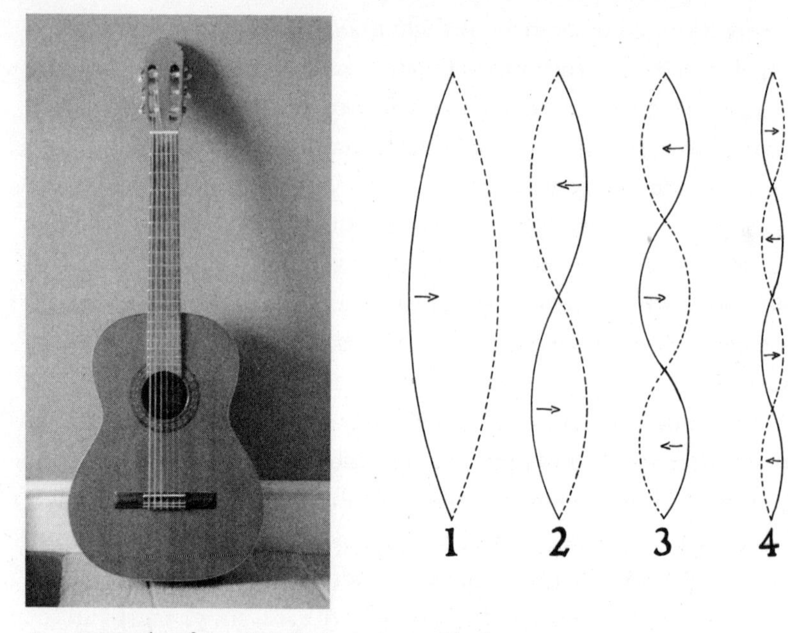

Some examples of ways in which a string could vibrate:
1. The whole string vibrates at 110Hz;
2. The string divides itself up into two halves which vibrate at 220Hz;
3. The string divides itself up into three thirds which vibrate at 330Hz;
4. The string divides itself up into four quarters which vibrate at 440Hz.
All these vibrations (with lots of others) happen at the same time, as a complex dance which repeats a whole cycle at the lowest frequency involved—110Hz.

drawings of a few of the individual ways that a string could vibrate from side to side without the ends moving.

Keeping the ends of the strings stationary automatically means that, to create the various vibration types (1–4 in the illustration above, and others), the movement pattern must divide the string up into one part or two parts or three parts—or any other whole number of parts—but not in any more complicated ways. You can't divide it up into four and a half parts, for example, because that would involve one of the ends waggling about.

So, while it is vibrating to and fro, the string does not just act as one long string swinging backward and forward at the fundamental frequency. It is involved in a complicated, wiggling dance which encompasses lots of vibrations happening at the same time—in the same way that the rapid, small wiggles of the swing chains were superimposed on the overall gentle swinging of the chain. The "whole string" movement of the guitar string will be accompanied by some "half string," "third string" and "quarter string" vibrations (and others).

Short strings vibrate at higher frequencies than long strings (if they are the same type of string under the same amount of tension). In fact, there is a direct relationship between the length of a string and the frequency at which it vibrates. If you halve the length of your string you double the frequency or, if you use a string which is one sixth as long as your original, then you multiply the frequency by six.

All the "shorter strings" we get when the string divides itself up vibrate at frequencies appropriate to their length—so the halves vibrate at twice 110Hz and the thirds vibrate at three times 110Hz and so on. Although the halves, thirds and quarters are all doing their own thing, like dance partners at a formal dance, they will all "return to base" to restart the dance at regular intervals. The overall pattern of the dance will repeat at the same rate as the lowest frequency involved, 110Hz in this case. This is why we call the lowest frequency the fundamental frequency of the note. This is the pitch of the overall note we hear—which is why we refer only to the lowest frequency when we are discussing musical notes. All the other frequencies join in to support the fundamental (rather like back-up singers), and this produces a richer sound.

What actually happens when you let go of the string is that all the types of vibration which involve lots of movement of the string near the plucking position begin at once. For example, if you pluck the string in the middle, you will get lots of the fundamental frequency and the "three times" frequency because both of these types of vibration involve lots of movement in the middle of the string (as you can see in the next illustration). You will not, however, get any of the "double" or "four times" frequencies because they require the string *not* to move in the middle.

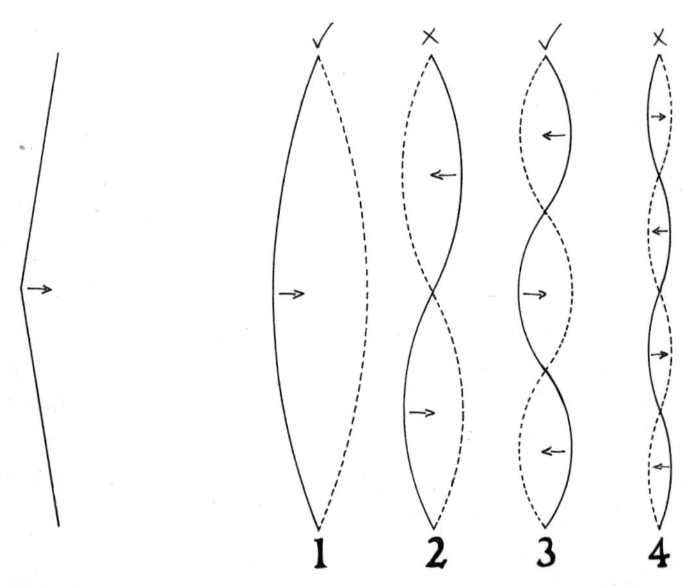

If you pluck the string in the middle, all the vibrations which involve the string moving in the middle can join in the movement. But vibrations which involve the string not moving in the middle will not take part. Here you can see the string (on the left) is being plucked in the middle—so only vibration patterns which allow movement in the middle of the string (ticked) are allowed to join in when the string begins to vibrate.

As another example, if you pluck the string one third of its length from one end, you will get lots of the fundamental, the "double" and the "four times" frequencies, but none of the "three times" frequency, because the "three times" frequency can't join in if the string is moving at that point, as you can see in this next drawing.

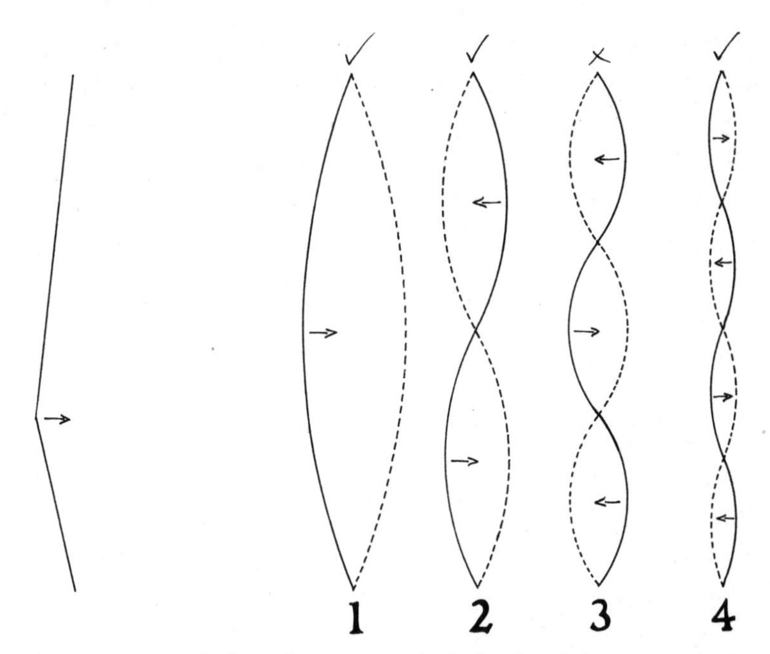

In this case, we are plucking the string one third of its length from one end. Once again, the vibration types which involve movement at this point (ticked) are allowed to join in, but the ones which require the string to be stationary at this position will not take part.

Now we get the radiant light of enlightenment shining down upon us. Plucking the string means we are going to get lots of different vibrations at once, and where we pluck it influences the various proportions of these vibrations in the mix. This explains why guitar strings sound different if you pluck them near the middle or close to one end. Wherever you pluck the string, the fundamental frequency will remain the same, but it will be supported by different proportions of the other vibrations. We have changed the pattern of the formal dance without affecting how often it completes a full cycle of movement. To get a new note, you must either choose another string, or shorten the one you are on by using the frets* on the guitar neck. Then you get a different fundamental, with its own group of related frequencies.

There is, of course, one group of musical instruments whose sounds don't involve collaborative harmonic mixtures—they produce noises rather than notes. These are the untuned percussion instruments such as cymbals, gongs and bass drums. They are among the most ancient instruments of all and date back to way before the introduction of table manners. Back in the good old days, people would bang along to the tribal dances on anything that came to hand. Nowadays things are a little more sophisticated, and using the skulls of enemies or close relatives is frowned upon, even among rugby players. Rock, pop and jazz bands use drums and cymbals almost continuously to provide a rhythmic drive to the music and this is the main reason why these instruments need to make a noise, not a note. If the drummer banged along using one or two notes all the time, those notes would dominate the tune and occasionally clash with the harmonies of the song. A repeated noise—a "thud" from a drum or a "tschhh" from a cymbal—provides rhythmic information without hijacking the music.

Drums are circular and rather like short columns, so they have a tendency to produce notes if you don't dissuade them from doing so (some drums, like orchestral tympani, are deliberately tuned to produce notes). A bass drum has two skins, one at either end of the "column," and the way to prevent it producing notes is to tune the skins to different notes. The drummer hits one of the skins and this compresses the air inside the drum, which starts both of the drum skins moving. However, because the two skins cannot agree on a mutually supportive pattern of movement, the noise has no identifiable pitch. The lack of mutual support is also the reason why the sound dies away rapidly—which is also useful if you want to produce clear rhythms.

* The use of frets to shorten guitar strings is discussed on page 52.

The reason I chose the "A" string of a guitar for the example in this chapter is because its fundamental frequency is 110Hz—and it's dead easy to see that 220Hz is twice that number and 330Hz is three times as big. But this book is about all the notes we use, so we can't keep on using this one example. The general rule is that any note is made up of a fundamental frequency together with its "twice" frequency, its "three times" frequency, its "four times" frequency, and so on. All these frequencies are called the *harmonics* of the note. The fundamental frequency is the first harmonic, the "twice" frequency is the second harmonic and the "three times" frequency is called the third harmonic—and so on for all the other numbers.

Although we have only discussed guitar strings so far, these principles are true of all musical notes—they all involve a family of vibrations linked by whole number relationships—and different mixes of these vibrations give the overall note a different character without changing its fundamental frequency.

The principle that strings produce different harmonic mixtures depending on where you disturb them also works for instruments where the string is bowed rather than plucked, such as violins and cellos. It provides a very useful method of controlling the sound of your instrument if you are playing a piece which repeats itself (as most pieces do). You can play the tune at first by plucking or bowing toward the middle of the string, which will give you a nice mellow tone. When you are repeating the tune later, you can get the same notes with a much harsher tone if you pluck or bow closer to the end of the string. This is a good way to enhance the musical interest for the listener and to increase or relax the tension of the piece. The guitarist can start off with a soulful, comforting tone as the singer tells his girlfriend that he senses that she is not entirely happy. Then later, the guitar tone can be adjusted to sound a damn sight less relaxed about the whole thing when he's singing the bit about how she ran off with that podiatrist from Schenectady.

The pressure ripples we saw on page 23 look rather complicated, but the vibrational patterns of the strings you see in the illustrations above are very simple. You might wonder how a collection of simple things can produce a very complicated one. In fact, any one of the individual string vibration patterns would produce simple pressure ripples in the air (like the ones on the left in the illustration below) if they acted alone, but in practice they never do so because we hear them joined together in groups. A real note might, for example, involve a lot of the fundamental and the third harmonic with smaller contributions from the second, fourth and ninth. The different mixtures of these various ingredients can give you an incredibly wide range of combined ripple shapes. The first person to realize that you could construct almost any repeating pattern out of combinations of simple waves was a Frenchman called Joseph Fourier, who was one of Napoleon's top men in the three closely linked intellectual fields of Egyptology, math and

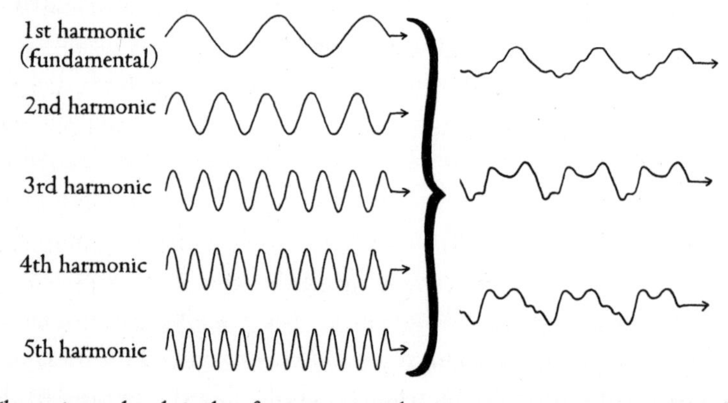

The various related ripples of a note can combine in various proportions to give different overall ripple shapes with the same fundamental frequency. The proportions depend on the instrument involved or on different ways of playing a particular instrument—you get the same note but a different sound (smooth, hard, etc.). The combined ripples on the right are the patterns for the flute, clarinet and guitar we saw earlier.

swamp drainage. Using spoon-bendingly complicated math, he managed to produce just about any repeating pattern you could imagine from these simple ingredients. Thankfully, there is no need for us to go into all that here. It is enough for us to know that if you put any musical ripple pattern into a computer it could work out the mix involved: "this much fundamental and that much second harmonic—with a smidgin of seventh harmonic and a twist of nineteenth" or whatever was needed to get that particular overall ripple shape.

So now we have answers to our three questions:

1. A string vibrates at lots of frequencies at the same time because the shape it started from allows lots of different frequencies to join in the overall dance of movement.
2. The frequencies are all related by whole number ratios because the string can only divide up its length into whole number divisions while it vibrates (because the ends of the strings can't move). The "half-length" strings vibrate at twice the fundamental frequency and the "third length" at three times, etc.
3. We hear an overriding fundamental frequency because we are hearing the string completing one entire cycle of its complicated dance at this frequency.

If you break a plate or rustle a paper bag, you hear a noise which is made up of a lot of unrelated frequencies with no repeating pattern—but all musical notes are made up of repeating patterns. Our brains can rapidly identify a sound as being made up of either a repeating or non-repeating ripple pattern and that is how we distinguish between notes and noises.

We know from experience that the same notes played on different instruments don't sound the same—even if they are the same frequency. For example, the 110Hz note produced when we twang a guitar "A" string sounds different from the 110Hz "A" note on a trombone. This is because the trombone note has different proportions of the various harmonics adding to the mix. These different mixtures are responsible for the distinctive sounds of various instruments—and this is the subject of the next chapter.

As well as controlling the sounds of different instruments, harmonics also influence our choices of notes which sound good together in harmonies and melodies. If, for example, we hear a note with a fundamental frequency of 110Hz played at the same time as one with twice this fundamental frequency (220Hz), they sound great together. The 110Hz note is made up of harmonics with frequencies of 110, 220, 330, 440, 550, 660Hz, etc., and the 220Hz note has harmonics of 220, 440, 660, etc. The reason the notes sound good together is because the frequencies of their harmonics have a lot in common—the frequencies of the harmonics of the higher note are the same as some of the harmonics of the lower note. Notes like this—where the upper note has twice the fundamental frequency of the lower one—are said to be an *octave* apart. This interval, the octave, is the cornerstone of all musical systems and we will be hearing a lot about it later in the book. Notes an octave apart are so closely related that they are given the same name. We know that the note with a fundamental frequency of 110Hz is an "A" but the note with twice this fundamental frequency (220Hz) is also an "A"—and the note with twice *that* fundamental frequency (440Hz) is also an "A," and so on. In fact there are eight "A"s on a piano keyboard—and to distinguish them from each other we number them (as you can see on page 8). Our old favorite, with a fundamental frequency of 110Hz, is called "A_2" and the "A" above it, with a fundamental frequency of 220Hz is called "A_3."

4. Xylophones and Saxophones: Same Notes but Different Sounds

Imagine, if you will, that nightmare scenario—a room full of musicians. They are in a frisky mood as twilight approaches, because it will soon be feeding time. To amuse themselves they are taking turns to play a simple tune—first the violin, then the flute, then the saxophone. Each instrument plays exactly the same notes as the one before but, of course, they all sound different—any listener could tell which instrument is which.

The distinctive sound of each instrument is called its *timbre* (pronounced "tarmbruh"). If we ask a sax player to play "Three Blind Mice," and then ask a xylophone player to repeat the tune (using the same notes), it will be obvious that the difference between the timbres of the instruments is enormous, so how can we say that they are playing the "same" notes?

To answer this question we have to think about what is important as far as our hearing is concerned. The main job of our hearing system is to keep us alive and so the first thing your brain and ear must do when they encounter a sound is to analyze whether or not it is a danger message. When analyzing a sound for its danger content, our brains concentrate primarily on the timbre of the noise, seeking to work out whether the sound is being made by a small animal (rabbit is on the menu) or a tiger (I am on the menu). Fortunately it doesn't take long for our finely tuned intellects to work out that we are unlikely to be eaten by a xylophone. The second most important thing to do is to work out which direction the sound is coming from. Once again the brain leaps into action: "That plinking noise is coming from over there—from the general vicinity of that xylophone."

Having worked out that the situation is musical rather than lethal, the brain concentrates on the frequencies of the notes being produced and their general arrangement into melodies and harmonies. In the context of music, the timbres of the notes have some importance, but this is secondary to the frequencies (or pitches).

We identify two notes as being similar if their fundamental frequencies are identical, irrespective of any difference in their timbre. The timbre adds extra interest to the situation—in the same way that shading adds information to an outline drawing. This musical shading can have a big impact on the emotional feel of the music, which is why those violinists who walk from table to table in Italian restaurants are unlikely to be replaced by xylophone players.

Although we are going to concentrate on the timbre of individual instruments in this chapter, it is worth remembering that, in many cases, the overall timbre of a piece comes from the combination of instruments involved. When writing a big orchestral piece, the composer spends a lot of time deciding which instrument, or combination of instruments, gets to play which bits of the melody and harmony, in order to present the music in the most effective way. It's quite a balancing act to combine the various loudnesses and timbres of the individual instruments to produce an overall "voice," or timbre.

In the case of a four-piece rock band, the distribution of notes is pretty obvious, but the instrumentalists can choose a wide range of timbres for their instruments. The lead guitarist will press various buttons during the course of the song to make the sound more or less aggressive, and the keyboard players can do the same, or move from instrument to instrument. I will describe this button-pressing malarkey in more detail at the end of chapter 5 in the section on synthesizers, but for now, let's look at the timbres of individual, non-electronic instruments.

Here are three notes of the same frequency with different timbres. These wave patterns represent the variation in air pressure experienced by an ear. Imagine these wave patterns "washing up" against the eardrum just like different types of waves washing up on a beach.

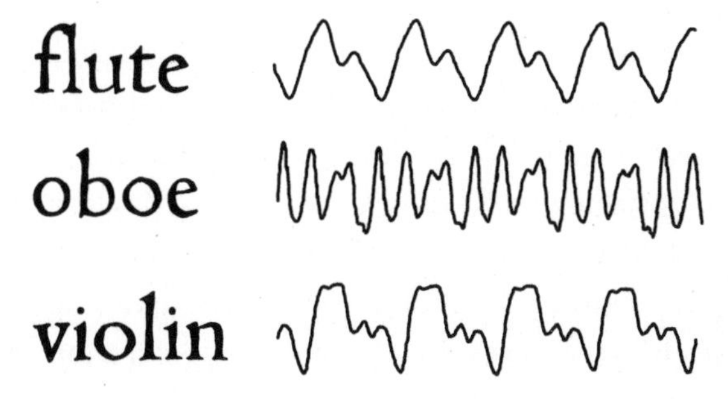

Recorded wave patterns of three notes with the same frequency but different timbres. Note that in every case there is more than one "hump" in each complete cycle. (Source: Measured Tones *by I. Johnston (Taylor & Francis, 2002)).*

In the case of the top wave pattern, the eardrum will move backward and forward in an even, regular way as the air pressure goes up and down. This will result in the brain experiencing a rather pure sound. The wave pattern shown here is a trace of a note from a flute—which sounds smooth and even to our ears.

The middle and lower waves in this illustration are also repeating patterns, but in this case the pressure variations felt by the eardrum are more complicated and jerky, causing the brain to experience this sound as richer and less smooth. These notes are the same fundamental frequency as the one played by the flute—and are therefore the same note—but this time they are played on an oboe in one case and a violin in the other.

But why should a flute make a sound which is smoother and less complex than that of a violin or oboe? To answer this question we have to think about musical instruments as machines which produce notes. All these machines are designed to produce repeating ripple patterns of pressure in the air and they all do this in different ways. For example, playing a flute involves a straightforward method of setting up vibrations in a column of air. There are no moving parts inside a flute, just this simple vibrating body of air. Playing a violin, on the other hand, involves a rather complicated method of vibrating a string by scraping it with a bundle of sticky horse hair (more about this later). The string then passes its rather jerky vibration onto the body of the violin—which is an unusually shaped wooden box. Although the overall vibration of the box will repeat at the fundamental frequency, various parts of the box will vibrate in different directions. So, rather than singing with a single voice (like the flute), a violin is more like a choir of several different voices, all singing the same note. Some of these voices are gruff, some are squeaky and, combined, they give a complicated, rich sound. The relative importance of the various parts of the "choir" changes as we play higher notes, where, for example, the squeakier members might have more influence, so the timbre of a violin differs quite a lot over its range. A skilled player can even play a single note with different timbres. If you play with your bow near the center of the string, you encourage the mellow part of your choir to contribute more to the note. If, on the other hand, you use the bow near the end of the string, close to the bridge, you will get a much harsher, more aggressive sound. On instruments which sing with far fewer "choir members"—like the flute—the range of timbres is much more limited, but even in these cases the timbre is different between high and low notes.

So, if instruments don't have a steady, identifiable wave pattern over their whole range, how do we recognize them so easily no matter what notes they are playing? Well, we make our decision about what type of instrument it is from two main sources of information:

1. The sound the instrument makes when the note is just starting;
2. The sound the instrument makes while the note is playing.

Let's look at these two things separately.

Differences between instruments when the note is just starting

It may sound daft, but we get a lot of our information about which instrument is playing from the unmusical noises the instrument makes just before each note starts, rather than from the notes themselves. Lots of experiments have been carried out to prove this—usually by playing recordings of slow music with the very beginning of each note removed. If this is done, it is difficult to identify what type of instrument is playing—even though the majority of each note is still on the recording.

There are lots of ways in which notes are started on different instruments. Before the notes sound out properly, the non-musical start-up noises made by plucking, bowing, hitting, etc., are easily identified by the human ear, even if the notes are very short and fast. This hearing skill is undoubtedly linked to survival—after all, if you can't recognize the "twang" of a bow and arrow very quickly you're not going to last long. By the way, these start-up noises are known as *transients*.

In the context of music we also identify instruments from the rise and fall in the loudness of the notes during their individual lifetimes. For example, a piano note starts suddenly and then fades off. A clarinet note, on the other hand, has a slower build-up and can remain at the same volume for several seconds. This volume variation of the note is known as its *envelope*.

Differences between instruments when the note is playing

You will remember from chapter 3 that a musical note is made up of lots of frequencies all sounding at the same time: the fundamental frequency, along with its other harmonics.

At the risk of being accused of scrimping on the illustration budget here, I would like to show you the final illustration from the previous chapter again because it shows us the basic principle behind the production of different timbres.

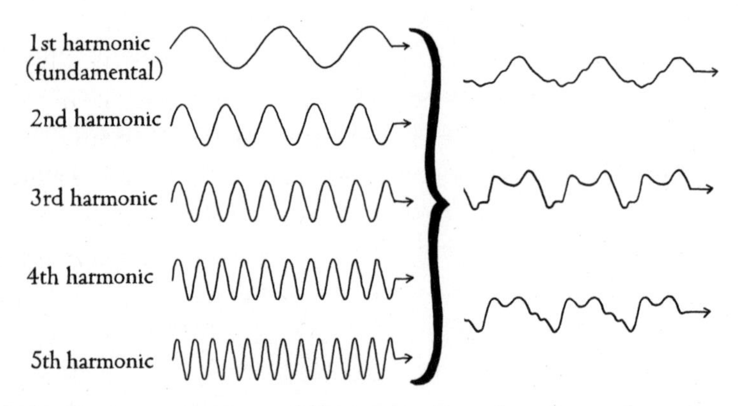

Harmonics can join together in different mixtures—depending on the instrument and how it is being played—to give different timbres.

Here we see the first five harmonics involved in a note joining together in different mixtures to produce notes with different timbres. There are many more frequencies involved in most real notes so these ripple patterns can become very complex. The reason why instruments have different timbres is because they produce notes which contain different mixes of these harmonics. For example, on a violin, the mixture of harmonics for the note middle C involves lots of fundamental frequency backed up by the second, fourth and eighth harmonics. On a flute, however, the same note involves mostly the second harmonic backed up by the fundamental and the third harmonic. In both cases lots of the other harmonics join in to enrich the sound.

As we saw earlier, the different mixes result in a much more complicated pressure ripple pattern for the violin than for the flute. As far as the physics is concerned, the flute ripple is closer to the shape you would get off a pure "fundamental frequency only" wave and so we could say that the flute produces a "purer" note. The odd thing is that, as listeners, we don't seem to favor purity over impurity. We enjoy the complicated sounds of the violin and saxophone just as much as the purer timbres of the flute, harp and xylophone. This is also true of our appreciation of singers. We like the purity of sound we get from Charlotte Church singing her version of "Silent Night" just as much as we appreciate the whiskey and smoke sound of Louis Armstrong singing "What a Wonderful World."

As I said earlier, the reason why the instruments produce different mixes of frequencies is because they are different shapes and sizes, and also because they make their sounds in different ways.

Let's take a closer look at the violin. Whatever note is being played, the sound is being produced by a vibrating wooden box which has a particular shape and size. The size and shape of the box makes it very responsive to certain frequencies and less so to others. Every note played on the instrument is made up of lots of related frequencies and, whatever note is being played, some of the frequencies involved will be among those "favored" by the shape and size of the box. As you might expect, these favored frequencies are produced a little more loudly than the others which make up the note. The technical name for the collection of favored frequencies of an instrument is its *formant*.

To understand what a formant is, let's consider a couple of notes played one after another on a pathetically bad cello. A real cello has a formant which favors lots of different frequencies, so that it sounds good for a wide range of notes. But we are going to invent a dreadful instrument which favors only a narrow range of frequencies. We will play the notes A and D on the instrument and assume that, although it vibrates at all frequencies to some extent, it favors only the frequencies close to 440 vibrations per second (440Hz).

When the A is played, we will hear its basic, or fundamental, frequency (110Hz), together with twice that frequency (220Hz) and three times (330Hz) and four times (440Hz), etc. On an instrument which had a uniform, fair response to all frequencies, we would hear an evenly mixed combination of all these vibrations. But, as I say, real instruments are not fair, they have favorites and in this discussion our instrument is entirely unreasonable and only favors frequencies around 440Hz, so the fourth harmonic (440Hz) will have more than its fair share in the sound we hear.

When we change the note to D, the frequencies we will hear will be the fundamental (146.8Hz), together with twice that frequency (293.6Hz), three times (440.4Hz), four times (587.2Hz), and so on. But the instrument hasn't changed: it still favors frequencies around 440Hz. So when D is played, our instrument will now favor the third harmonic frequency of 440.4Hz and we will hear that component of the note more prominently than we would from an unbiased instrument.

These more prominent harmonics do not affect the fundamental frequency of the note. They just change the mix of harmonics, which affects the timbre of the note we hear—so these two notes would have different timbres. An instrument like the pathetic cello I have just described would be useless, because it has only one favored frequency, which would dominate any music you played on it. Fortunately, real instruments have lots of favorite frequencies and this ensures that all the notes are produced clearly. However, each instrument type has its own family of favored frequencies and this is one of the reasons that cellos sound different from violins even if they are playing the same note.

It is also interesting to note that, in many cases, the ripple pattern of a note changes during its lifetime. The illustration below shows this very clearly—the ripple traces are from near the beginning, middle and end of a single note played on a harpsichord. These changes are different for each type of instrument and add to our assessment of the timbre of the instrument involved.

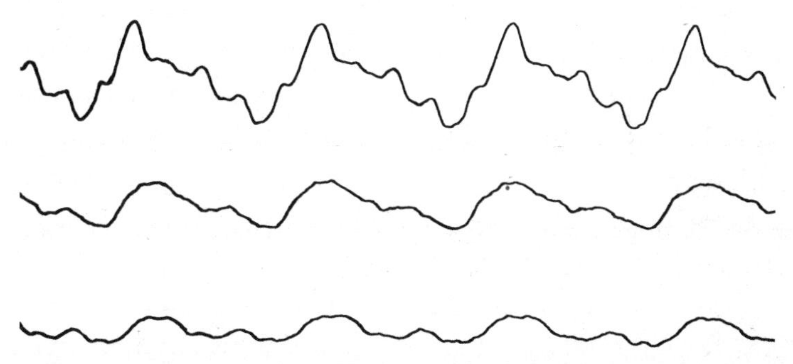

These three ripple patterns show how a note changes during its own lifetime. Here we see traces taken from near the beginning, middle and end of a single note played on a harpsichord. (Source: C. Taylor, *Exploring Music,* Taylor & Francis, 1992).

Now that we understand the basics of timbre, it's a good time to compare the ways in which various instruments produce the sounds they make. The following chapter investigates the musical personalities of some of our favorite instruments, from violins to synthesizers.

5. Instrumental Break

Note production on different types of instruments

It's all well and good for me to speak of flutes as having "a simple shape" and violin strings as having "rather jerky vibration," but to achieve a deeper understanding about how instruments produce their various timbres we need to look at a few instruments in more detail. We don't need to look at all of them because many of them share the same principles. Clarinets, for example, are very similar to saxophones, and cellos are effectively just big violins.

I have chosen three stringed instruments: the harp, the guitar and the violin, because they each have something to tell us about the vibration of strings. The harp is the simplest to explain because the strings are simply plucked. The guitar involves both plucking and shortening the strings to get the notes you want, and the violin introduces us to the concept of bowing rather than plucking.

We then move on to three instruments which produce notes from the action of wind in tubes:* the organ, the penny whistle and the clarinet. The organ provides a good introduction to how notes are produced by individual tubes, and the penny whistle is used to explain how we can get several notes from a tube with

* Traditionally the organ is referred to as a keyboard instrument and the thought-police will take you away in handcuffs if you call it a wind instrument. So I'm calling the second three instruments "wind in tubes"—let's all hope I get away with it.

holes in it. The clarinet is a typical example of how reeds are used to get a rich, complex timbre from a tube of air.

After these wind-activated instruments we move on to two tuned percussion instruments: the glockenspiel and the piano; and I will finish off with a few notes about synthesizers. Between them all these instruments have a lot to tell us about timbre, but the following descriptions also contain a lot of other information about the basics of note production.

Three stringed instruments

The harp

A harp is basically a set of strings which are stretched between a solid beam of wood and a hollow wooden box (harpists might feel a little hurt by this coarse description of their instrument but, as no one has ever been beaten to death by a harpist, I feel bold and unafraid). As I described in chapter 3, when we pluck a string we stretch it in one direction and then let it go. The string then tries to return to its original, straight condition, but it keeps overshooting until after a while it runs out of energy and becomes straight again. A stretched string by itself will not make much sound, but if you attach it to a hollow wooden box (a guitar, violin or harp, etc.) the vibration is passed on to the air much more effectively and we hear a louder note. A harp note is loud at first and then dies away as the movement of the string diminishes, which gives us that distinctive "twanged string" sound that we also get from guitars.

The way we excite the string couldn't be much simpler, and the soundboard is basically a flat piece of wood which forms the top of an uncomplicated box shape. The timbre of a harp therefore tends to be a very pure, sweet note. This sweetness of nature means that harps are chosen to do certain musical "jobs" in the orchestra but not others. One famous example of harp use is the slow movement (adagietto) of Mahler's Fifth Symphony. This piece, which has been used in several films, consists of about ten minutes of lovelorn strings with a slow harp accompaniment. The harp doesn't seem to be doing much but it adds a lot of magic to the piece. In particular, the "loud at first then dying away" character of each harp note helps to add rhythm and a feeling of forward motion to the long notes played by the strings.

The fundamental frequency of the vibration of the plucked string (and therefore the note we hear) is determined by three things:

1. How tightly the string is stretched (a tightly stretched string is in much more of a hurry to return to its straight condition than a slacker string and therefore moves to and fro more rapidly—which gives us a higher frequency and therefore a higher pitched note).
2. The material the string is made of (dense materials such as steel move more slowly to and fro than light materials such as nylon—so denser materials give lower notes).
3. How long the string is (longer strings give lower notes because vibration information needs to travel up and down the strings, and the journey takes more time on longer strings).

On a harp, or any other stringed instrument, we need all the strings to be pulled pretty tight or we won't get clear notes (a thin, slack string can produce the same frequency note as a thicker, taut one but the taut one will be clearer and louder). For this reason we design our harps with strings of different lengths, made of light or dense materials in order to get a wide range of notes from a set of tightly stretched strings. The lower note strings are made of steel and the higher notes are made of nylon. A range of different string lengths is given automatically by the (approximately) triangular shape of the instrument.

The only adjustment to the timbre of a harp comes as a result of where you pluck the string. If you pluck the string somewhere near the middle, then the note is at its most simple and you get the smoothest timbre you can get from any stringed instrument. If you want a harsher sound, then you need to pluck the string near one of its ends. Wherever you pluck the string along its length has an effect on the mix of the fundamental frequency with its various harmonics—just like the guitar string we discussed in chapter 3.

The guitar

The guitar is another instrument which involves plucked strings attached to a hollow box. Each guitar string is held tight between the *nut* at the end of the neck and the *bridge*, which is positioned on the body of the instrument. When you pluck a string you get a harp-like note. Most guitars, however, only have six strings and obviously we need more than six notes—otherwise guitarists would have zero sex appeal.

There are, in fact, more than forty notes available on any guitar and these are produced by pressing the strings against the neck of the instrument to shorten them. There are bits of wire called "frets" embedded in the neck and, when you press the string against the neck, it is held tight between the nearest fret and the bridge. Because the string is held between these hard objects it gives a clear, harp-like note when plucked (if you just held the string against a neck without frets your soft fingertip would soon absorb the vibration—and the resulting note would be more of a "thunk" than a "ding").

Because of the way guitars are designed it is possible to hold all six strings down at the same time and twang them all together, as

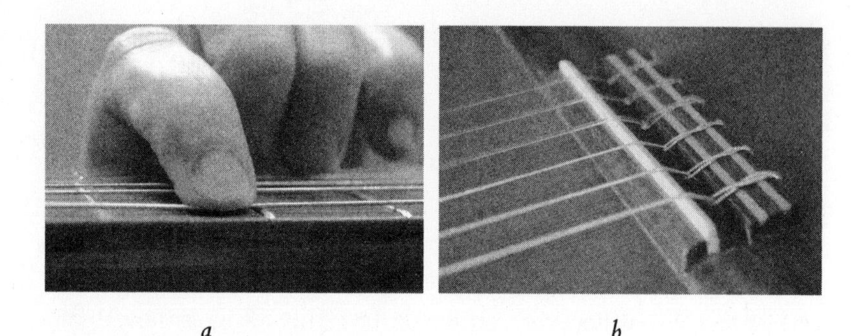

a b

Shortening a guitar string by the use of frets: the string is held between the chosen fret at one end, and the bridge of the guitar at the other.

you can see in the picture below. This ability to play chords (several related notes at the same time) is one of the reasons that guitars are popular for accompanying songs. Expert players can play chords and tunes at the same time and classical or jazz guitarists are often required to play two or three tunes simultaneously.

For obvious reasons the timbres of the harp and the guitar have a lot in common—they both involve plucked strings. If you are

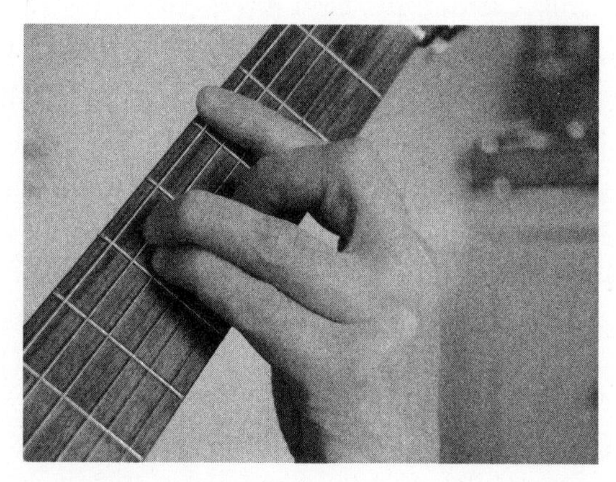

It is possible to shorten several guitar strings at the same time to produce chords of related notes. Here, all six strings are being shortened to produce a combination of notes which make up a major chord.

familiar with either instrument it is usually quite easy to spot the difference between their sounds because each can do things the other can't—for example, only a harp can produce those "zinging" scales and only a guitar can produce rapid, scrubbed chords.

If you were listening to one of these instruments on the radio, you would be able to pick up a lot of timbre information from how the notes end. Generally, guitar notes end more suddenly than harp notes because the guitarist often has to use the same string for the next note in the tune. Harpists, on the other hand, have one string per note and their technique often involves playing the next note before the previous one has finished. This gives us a gentle fading away of overlapping notes.

Another difference between the two instruments is that a guitarist can wiggle his finger on a string to stretch it slightly to and fro over the fret. This wiggling changes the tension on the string a little, and makes the pitch of the note go up and down a few times a second. The effect is called *vibrato* and is used to give notes (particularly long notes) a trembling, "romantic" depth. The effect can be achieved on quite a few instruments, especially violins, violas and cellos, and is sometimes used (and occasionally overused) by singers. You can hear different degrees of the use of vibrato by singers if you listen to various recordings of the jazz song "Cry Me a River." Rock and blues guitar players are also keen on vibrato, particularly on long notes in solos. To hear vibrato on a classical guitar, I recommend John Williams or Julian Bream playing Prelude No. 4 by Heitor Villa-Lobos.

The violin

The violin is, once again, an instrument with stretched strings attached to a hollow box. In this case we have only four strings and, as is the case for the guitar, different notes are produced by pressing the strings against the neck of the instrument to shorten them. On a violin, however, there are no frets. As I said above, this means that if you pluck a held-down string you get a "thunk" rather than a clear note. This thunking noise is called *pizzicato* and it's occasionally used by composers to get a tuneful but percussive effect from violins and other stringed instruments—the Pizzicato Polka by Johann Strauss is a great example of this.

Fortunately for all involved, violin players are only rarely called upon to pluck their strings. Their usual method of supplying the string with energy is rather more complicated and it allows the instrument to produce clear, singing notes rather than thunks.

To produce long, singing notes from a violin (or viola, cello or double bass), you need a bow. A bow is basically a collection of hairs from a horse's tail which are held taut by a specially shaped stick. The stretched horse hairs are made to be slightly sticky (but dry) by rubbing them with a material called rosin. Rosin is dried-out resin—that sticky, gluey stuff you sometimes see on pine trees.

A close-up photo of a violin bow showing the band of horse hairs, which is drawn across the string to make it vibrate (the horse hairs are usually bleached white—as they are here).

Rosin manufacturers collect the resin from pine trees and dry it out into little blocks to sell to violinists and other string players.

Before I explain how the bow works, I would like you to do something for me: put this book down and go over to the nearest window, computer screen or TV. Now lick your fingertip and rub it backward and forward across the glass (this doesn't work as well on plastic laptop screens). Within a couple of seconds you should be able to produce a squeaking noise. This noise is caused by the stick–slip motion of your fingertip across the glass surface. Stick–slip motion is just what it sounds like: motion made up of alternately sticking in one place and then slipping forward quickly before re-sticking and then slipping again.

The pressure of your finger against the glass tends to hold your finger in the same place but the forward pushing of your arm forces your finger to move—and your saliva helps to lubricate that movement. So, your finger starts off by being stationary and the forward pushing force builds up. Then, when the pushing force is enough, your fingertip moves quickly forward a fraction of a millimeter. This relaxes the forward pushing force and the fingertip can stop again—but then the pushing force builds up again to repeat the cycle (it repeats hundreds of times every second). In this way your fingertip alternately sticks and slips as it moves across the glass—and this produces the noise you hear.

As the sticky horse hairs of the bow are drawn across a violin string (see the illustration opposite), they undergo this slip–stick motion and this continuously excites the string—as if it were experiencing a tiny plucking action hundreds of times every second. In this case, the string is pushed to one side by the sticky bow, but once it is pushed far enough, it slips back to being straight and then overshoots (just as a plucked string will) before the stickiness of the bow grabs it again and pulls it back to where it started slipping from.

Drawing the bow across the violin string makes the string vibrate.

The frequency with which the slip–stick action happens is determined by the usual things which govern how often strings vibrate backward and forward: the tension the string is under, the material it is made from, and its length. The length of the string is changed by pressing the string against the neck of the instrument. As I said earlier, plucking a string held in this way (without frets) would just produce a short "thunk," but the action of the bow effectively re-plucks the string every time it vibrates backward and forward—and this produces the long singing note we associate with violins.

Bearing in mind the jerky way in which the string is vibrated and the complicated shape of the violin body, it is not surprising that the timbre of the violin is complicated and full of character.

It is worth pointing out another big difference between guitars and violins here; a guitar player (playing a properly tuned guitar) only has to put his fingertip down between two frets to get a note which is in tune with other instruments because the string length will be determined by the position of the fret involved. A violin player, on the other hand, can easily put her finger down in the wrong place on the neck and produce a random note which is not in tune with anything (I will be discussing these differences in more detail in the section on "Choosing an instrument" in chapter 11). For this reason it takes longer to become a competent violin player than a competent guitarist. This is also the reason why only good violinists play more than one note at a time, whereas a merely competent guitarist will find the playing of up to six notes at once quite straightforward. "Competent" here means someone who can play for five minutes at a wedding without having hors d'oeuvres thrown at them. Becoming a good violin player or guitarist takes about the same amount of time and effort because the demands of the two instruments differ—"good" in this case means that people will pay money to hear you play.

Musical training to get to the expert stage usually takes about ten years and continues for as long as you play the instrument. In fact, after lots of different investigations into skill acquisition, it is now generally accepted that it takes about 10,000 hours to achieve expert level in almost any activity—from landscape gardening to karate. Musicianship fits this model—so that's just over two and a half hours of practice a day for ten years. Of course, we're talking about professional levels of skill here; you can achieve a very satisfying level of musicianship if you just give it an hour a week.

Professional-level training usually pushes you toward the limits of what a human being can achieve on an instrument, but because instruments are designed differently, the demands on the musician are also different. I could teach you to play the melody to "Baa Baa Black Sheep" on a piano in fifteen minutes, but if you study the instrument to a "good" level it would be quite normal to expect you to play it while harmonizing the tune with lots of rippling accompaniment and chords involving six notes at a time—because playing just one note at a time on a piano is very easy. On other instruments such as the French horn or bassoon, it is difficult to produce even a simple tune reliably. We should all be particularly grateful to musicians who struggle through the early years of instruments like these, which are especially difficult for beginners. I couldn't do it—I found the early stages of classical guitar quite painful enough.... Come to think of it, my family found my early guitar twanging quite painful enough too.

Music from wind in tubes

The church organ

To describe how an entire church organ works would take too many pages—they really are incredible feats of engineering. All I want to do here is describe how the simplest type of organ pipe produces a note.

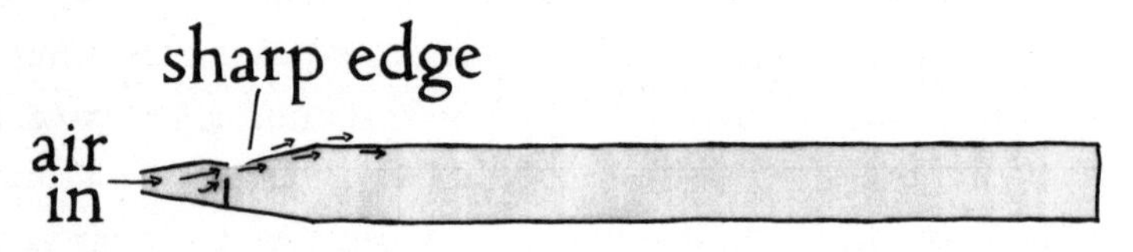

An organ pipe (closed-end type), showing the air flow over the sharp edge of the whistle.

The simplest church organ pipe is basically a tube with a whistle on one end that is closed at the other end. A whistle is just a container which forces a flow of air over a sharp edge or blade. In an organ pipe the whistle is the bit which generates the noise, but the frequency of the note produced is determined by how long the tube is. Let's look at how this works.

First of all we need a jet of air. In the good old days, starving children could be hired at very reasonable rates to operate mechanical bellows to produce the air needed by church organs. Nowadays, we usually use an electric air compressor. When you press one of the organ keys it opens a valve below one of the organ pipes and the air flows into the chamber at the bottom of the pipe. The air then escapes from this chamber through a narrow opening, as a fast-flowing stream or jet. This air jet then flows directly across the sharp edge shown in the previous illustration, and this sets up a vibration in the column of air in the tube. To understand how the note is produced we need to know what happens when a jet of air flows across a sharp edge.

When a jet of air hits a sharp edge it doesn't divide calmly into two streams. In fact, there is quite a lot of confusion at the edge and the air tends to alternate between one side of the edge and the other.

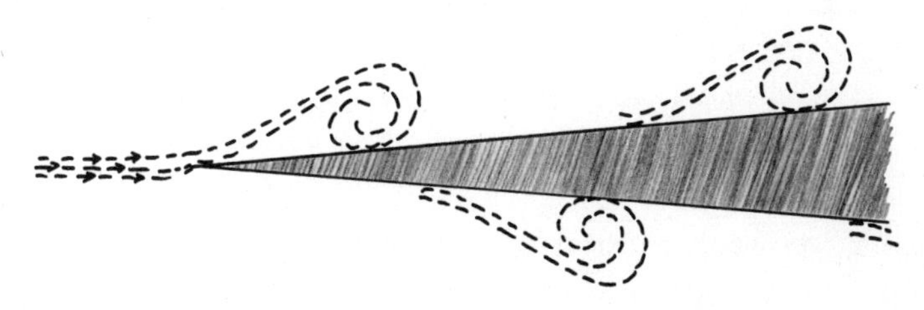

When a jet of air meets a sharp edge, the air "takes turns," going mostly to one side and then mostly to the other.

If this "taking turns" sounds unnatural and unlikely to you, then think of it this way: imagine road traffic coming to a long island in the middle of a one-way street. All the drivers are in a hurry so they will each take the route which looks most empty to them as they approach the island. So, Fred sees that the left-hand side of the island has fewer cars on it—and he will go that way. Jane, who is driving the next car, will see that, because Fred's car has gone left, the right-hand side of the road has fewer cars on it—so she goes toward the right. The "best way to go" swaps regularly from side to side and the cars take turns in going down the right-hand or left-hand route. The air from a jet finds itself in a similar situation when it meets a sharp edge and so the stream of air takes turns going first to one side and then the other. The deciding factor for the stream is the pressure on either side of the edge. Air, like any other gas, always tries to move to areas where the pressure is low in the same way that water always flows downhill. The air approaching the edge will "notice" that the pressure is higher on one side of the edge than the other and will head for the low-pressure side, but in going that way it increases the pressure on that side, so the next bit of air in the stream chooses the other side of the edge.

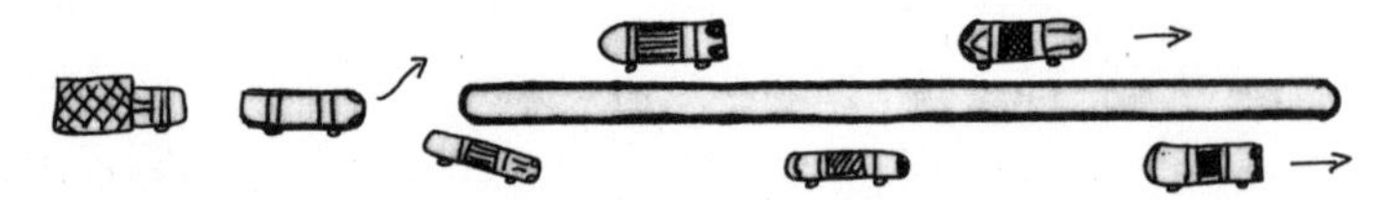

Traffic "taking turns" at a long traffic island in a one-way street.

In an air-powered musical instrument, the frequency of the "turn-taking" becomes linked to the length of the tube by a phenomenon known as *resonance*. Resonance is necessary for the production of any musical note and needs some explanation.

RESONANCE

Resonance is a process by which a small amount of effort repeated at the correct frequency can produce a big effect.

For example, if you are pushing a child on a swing, you can make the swing go very high with very little effort as long as you give a small push at exactly the right time—you need to push just at the moment the swing starts to move away from you.

A swing is like a pendulum, and the only thing which changes how quickly it can perform one complete backward and forward cycle of movement is the length of the chains or ropes attaching the seat to the frame. It doesn't matter how hard you push, how high they go, or how heavy the child is—one swing to and fro will always take the same amount of time. The only important thing is to match the frequency of your pushes with the natural rhythm of the swing and you will get a big effect for the minimum amount of effort. If you try any other frequency then things will go wrong. For example, if the swing takes three seconds to go and come back, you must push at three-second intervals. If you decide to be hot-headed and push every three and a half seconds, then the swing won't even be near you during some of your "pushes" and at some point in the near future you will push when the swing is approaching—and you'll have to spend the rest of the afternoon taking little Jasper to the emergency dentist.

If you would like another example of resonance which doesn't involve making a fool of yourself in a playground, try this. First of all you need to half-fill a fairly big container with water—you could use a bath or a sink. Now waggle your hand backward and forward rapidly in the water, keeping your hand flat (like a canoe paddle), as in the illustration below. The effect will be that of a little storm—lots of small disorganized waves—because you are fluttering your hand to and fro too fast to get any resonance. Now try moving your hand very slowly backward and forward. This time you will just get lots of little ripples because you are moving too infrequently to get any resonance. Finally, put some towels on the floor and get some resonance going. You only need to move your hand the same amount as before, but if you do it at the correct frequency you can get all the water moving backward and forward in one big wave. To achieve this, just push the water from left to right at different frequencies until you get an overall biggish wave going to and fro, and then follow the rhythm of that wave.

As with young Jasper's swing, you can't affect the natural rhythm of the wave—you must match that rhythm if you want to get the maximum effect for the minimum amount of effort. I've

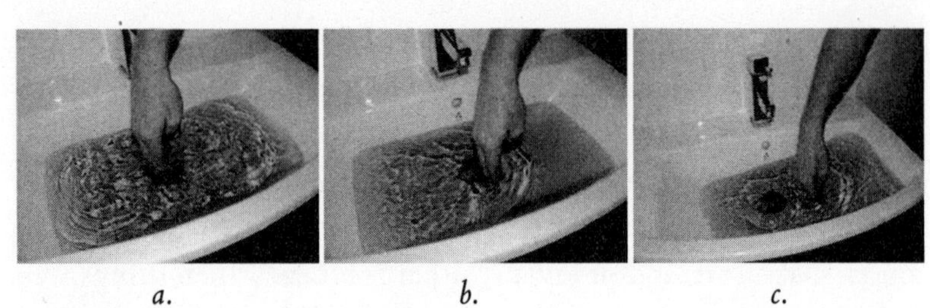

a. b. c.

Waggling your hand to and fro in a basin of water: a. too frequently (mini-storm); b. too slowly (ripples only); c. if you match the speed of your hand to the speed of a biggish wave, you will eventually get one big wave going, which will spill over the edge of your basin if you are not careful. This is a resonant effect, like pushing a swing. (For these photos I colored the water to make it more visible.)

just done this experiment in my bath and I found that the whole backward and forward cycle took about three seconds; in my sink it takes about one second. This is because my sink is about one third as long as my bath, so the wave only takes one third of the time to travel from one end to the other. This is an important point about resonant frequencies. Whether we are talking about water in a bath or air in a tube (like an organ pipe), the resonant frequency increases as the container gets shorter. There is an exact relationship between the two. If, for example, container "A" is one fifth the length of container "B," then the resonant frequency of container "A" will be five times that of container "B."

Resonance is the reason why some singers can smash wineglasses using their voices. When you tap a wineglass, the bowl of the glass bends inward (away from your finger) and outward hundreds of times a second in a repeating pattern and produces a note of a certain pitch. The glass is basically throbbing and producing pressure changes in the air. You can make this work backward if you sing the same note back at the glass loud enough. Instead of the bending glass producing the note, the note can cause the glass to bend. Glass isn't particularly flexible, so if you make it bend too far by singing very loudly, the glass will break. You have to sing exactly the note that the glass gives off if you tap it or resonance will not happen. The pressure waves of the note you sing will only "push the swing" at the right time if they arrive at the right frequency. Tapping the glass does not contribute toward breaking it—it's just a method of identifying this resonant frequency. Professional singers are best at shattering glasses, because they have been trained to recognize and reproduce pitches accurately and they are also trained to be loud—which means that the pressure changes of their notes are large. If you want to try your skills at glass vandalism, you should use a large, old, thin-walled glass. The glass needs to be large so that the note is low enough for us non-trained singers to reach. It needs to be thin-walled so that it's not very strong, and preferably old (and therefore covered in lots of little scratches, which will encourage it to break).

But perhaps we should leave your grandfather's priceless collection of wineglasses alone and get back to our organ pipe.

Each little puff of air inside the tube, created by the turn-taking at the sharp edge, travels along the tube as a wave of pressure. It then hits the closed end of the tube and bounces back toward the area near the sharp edge. After this has happened a few times, one of the returning waves will meet a newly created wave and the two of them will join forces bouncing up and down the tube. This bigger wave then sets up a resonance effect which controls how often the turns take place at the sharp edge. Turns will be taken at a frequency which is determined by how quickly the pressure wave can complete the round trip from the sharp edge to the end of the tube and back again (so—the longer the tube, the lower the frequency). This resonance effect begins to operate after only a fraction of a second and is the cause of the note we hear from the pipe. The note created by this effect will, of course, have the same frequency as the pressure wave bouncing up and down the tube.

Simple organ pipes come in two types—the ones we have just

discussed, which are closed at one end, and others which have their ends open. You might wonder how the wave can rebound off the end of the pipe if it is left open (I certainly did when I first heard about it). Well, the process is a little different from straightforward bouncing off a closed end, but the result is very similar and, once again, a resonant effect is built up to give us a note. As I said earlier, with a closed-end pipe a wave of high pressure rushes up to the end of the tube and bounces off it. If the end of the tube is open, the high-pressure wave leaves the tube and, as it does so, it leaves behind a low-pressure zone at the end of the tube. This results in a low-pressure wave which then rushes back down to the whistle end of the tube. All this rushing backward and forward sets up a resonance effect (similar to the simple bouncing off a closed end) and a note is produced.

A typical church organ is a big collection of individual whistles. The frequency of the note produced is determined by only two things: how long the tube is, and whether or not the end is closed (a closed tube produces the note an octave lower than an open-ended tube of the same length). The timbre of the note given off by a whistle can be affected by the shape of the tube in cross section (you can get round, square or even triangular ones), but one of the most important factors is the width of the tube. Thin tubes encourage high frequencies—so they will give you a mix involving less of the low-numbered harmonics and lots of contribution from the higher harmonics. A note with lots of high harmonics like this has a very bright—or even shrill—sound, whereas a note from a fatter tube will concentrate on the fundamental and its close companions—to give a more rounded sound.

Organ builders specialize in giving their instruments a wide range of timbres, so they fit them with lots of different sets of whistles. They might have one set of thin tubes, a set of fat tubes and several intermediate sets. By pulling various buttons, called *stops*, in or out, the organist can choose to play one set or the other. And that's not all; the organ builder will also add several sets of tubes which have completely different timbres, conical ones for example, and others with reeds, which make them sound like clarinets. This gives us lots of choice in the timbre, but the great thing is that groups of these different sets of tubes can be played at the same time to give you hundreds of possible combinations. You might, for example, use the thin clarinet-like ones together with the fat open-ended tubes, and then change to all the conical ones together. For a big finale you might want all the tubes on the organ to join in—which will require you to *pull out all the stops,* which is where that phrase came from.

By the way, those long shiny tubes you see on big church organs? I'm afraid they are just for decoration. The real tubes are hidden behind them.

The penny whistle

Unfortunately, penny whistles no longer live up to their name. The laws of economics have taken their toll over the years, and nowadays they should be called "500 penny" whistles if you want to be pedantic about it. In spite of this inflation, they are still the cheapest, most beginner-friendly instruments you can get hold of, and, in the hands of an expert, they sound wonderful. I was trying to learn "The Lonesome Boatman" on mine, and I intend to get right back to the project as soon as my neighbors drop their noise complaint.

The penny whistle is similar to an organ pipe, in that it is a tube with a whistle on one end. In this case, however, the tube has several holes drilled into it. You can change the length of the resonating tube by closing these holes with your fingers. With your fingers over all the holes, the resonating part of the tube is as long as the whole tube and you get the lowest frequency note. If you take one

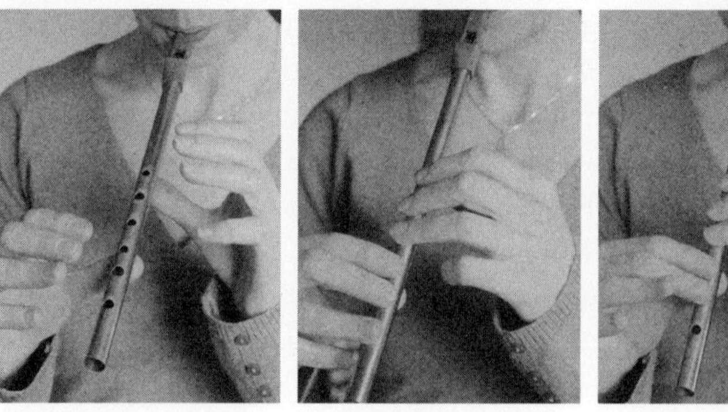

a. All holes open. *b. All holes closed.* *c. One hole open.*

Playing different notes on a penny whistle

finger off, the air in the tube only resonates up to the first hole it comes to (the one you have just taken your finger off). This means that the tube is now shorter and therefore the frequency of the note will increase. This effect of shortening the resonating length of the

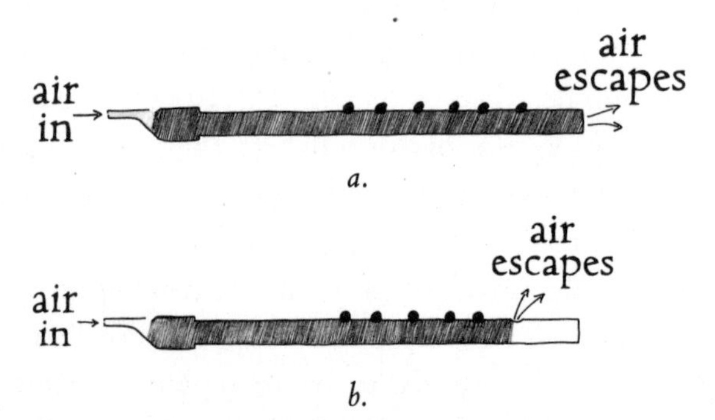

What happens inside a penny whistle (the shaded area shows the portion of the air in the whistle which is resonating to produce the note): a. with all the holes closed, the air resonates as far as the end of the tube and the lowest note is produced (long tubes = low notes); b. if some of your fingers are taken off the holes, the air only resonates as far as the first open hole (because the pressure waves can escape at that point). This "shorter" tube gives us a higher note.

tube by taking your fingers off the holes is demonstrated in the drawing opposite. The pressure waves bounce backward and forward only as far as the first place they can escape from—the nearest uncovered hole.

Penny whistles are designed to produce only the notes of a major scale. There are seven different notes in a major scale so we only need six holes (we get one note when none of the holes is closed and we get the other six by closing off the six holes with our fingers). One interesting thing about penny whistles (and other wind instruments) is that not all the holes are the same size. We could produce penny whistles with six identically sized holes—but they would be more difficult to play because some of the holes would be uncomfortably close together. To avoid this problem, it is possible to keep the same note and move the hole toward the mouthpiece as long as you use a smaller hole.

I said earlier that the pressure waves only bounce backward and forward as far as the first uncovered hole—but this is only true if the hole is big enough. If you use a smaller hole, then the pressure waves can be fooled into acting as if the tube is slightly longer than the distance from the mouthpiece to the hole—as shown in the next illustration. The sound pressure waves cannot fully escape from the small hole and so the next hole along is also involved, and the resonating effect stops somewhere between the two holes.

If the hole in the penny whistle is small, then the pressure waves cannot easily escape. In this case the resonance effect (shaded) carries on for a few millimeters after the first open hole because two holes are sharing the job of allowing the pressure waves to escape. The tube therefore gives a slightly lower note than would normally be expected from a hole in that position.

This principle is used by manufacturers of wind instruments to position holes in the best places for ease of playing. It is also used by advanced penny whistle players to get extra notes (between the ones in the major scale) by half covering holes—which has the effect of making the hole smaller. Real experts, like the over-talented folks who play "The Lonesome Boatman" on YouTube, can use this small hole/big hole thing to slide gradually from one note to another. They lift their finger slowly off the hole so that it gradually appears to get bigger as far as the air inside is concerned—and the length of the column of air which is resonating slides up and down from one hole position to another. Of course, I only understand this in theory—my attempts to do it sadly resulted in that unfortunate incident with the dog next door.

Recorders also make use of the small hole/big hole effect to allow the player to get two notes from one finger position. Recorders often have two small holes next to each other, as you can see in the illustration below. If only one of these holes is uncovered, you get the "small hole" note, and if both of these holes are uncovered, you get the "large hole" note—which is a semitone higher.

Double holes, like the ones on this recorder, allow the player to get two notes from one finger position—because you can have either a small hole (when one is uncovered) or a large hole (when both are uncovered).

More complicated pressure wave resonances can be set up in tubes with holes in them by closing up combinations of holes with open holes between them. This helps us to obtain the maximum number of notes from a limited number of holes.

Finally, it is possible to get resonant frequencies which are one or two octaves higher by blowing harder into the instrument.

The combination of all these note-producing techniques makes it possible to produce a surprisingly large number of notes from a penny whistle which has only six holes drilled into it—but it doesn't stop penny whistles from being very *very* irritating in the wrong hands. ("The wrong hands" in this context means "anybody else," of course.)

Concerning timbre, as I said in the discussion of organ pipes earlier, thin pipes make bright noises because they promote the higher numbered harmonics. These higher frequency family members are encouraged even further if the air speed is increased—as it is when you blow harder to get the upper octaves. This is why the upper notes on a penny whistle sound so shrill. Sorry, I need to reword that last sentence.... This is why the upper notes on a penny whistle sound so damned shrill.

The clarinet

To get a musical note out of a tube we need to have a situation where repeated puffs of high pressure are sent up the tube. At first the puffs can be rather disorganized—but very rapidly a resonance is set up and the frequency of the puffs becomes fixed, and a note is produced. We have seen that, in the case of an organ pipe or penny whistle, the puffs are produced by the "turn-taking" of a jet of air as it passes over a sharp edge. In the case of a clarinet, the jet of air is divided up into a stream of individual puffs by a reed at one end of the tube which is forced, by your breath, to flap open and closed hundreds of times every second.

The next illustration shows how the reed works in the clarinet mouthpiece. The clarinet player presses gently on the lower surface of the reed with her lower lip—this closes the path for any air to get into the tube of the clarinet. The player then blows fairly hard while slightly releasing the pressure on the reed. Eventually the air starts to squeeze through the tiny gap between the reed and the rest of the mouthpiece. A balance is then set up between the air forcing the reed open and the bottom lip squeezing it shut. The reed then opens and closes hundreds of times a second—and a stream of puffs of air is released into the tube of the clarinet. As in the case of the organ pipe and the penny whistle, the frequency of the puffs quickly becomes controlled by the distance between the mouthpiece and the first open hole in the tube (or combination of open holes).

Like the violin, the timbre of a clarinet is complex and full of character. The reason for this is that, in both cases, the way we give

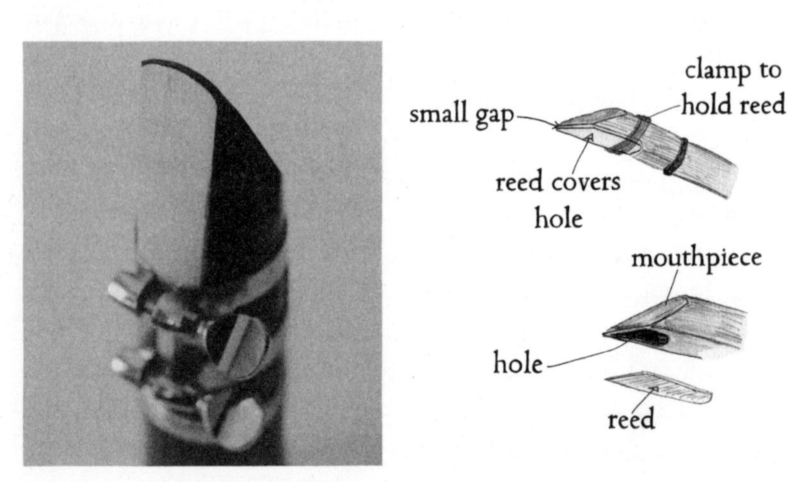

A clarinet mouthpiece. The clarinetist's bottom lip pushes upward on the reed, closing the small gap. The clarinetist then blows air through the gap. The pressure of the blown air opening the gap, and the lip pressure closing it, balance out, and the gap opens and closes hundreds of times every second to produce musical notes.

energy to the instrument involves a frequently interrupted action. Instruments which produce a smooth timbre—like harps, guitars and flutes—involve a vibration which swings to and fro in a regular, even manner. We have seen that a plucked string does this and so does a jet of air when it meets a sharp edge. A violin string, however, is dragged relatively slowly in one direction by the bow and then slips rapidly back the other way before being grabbed by the bow again—so the vibration is slow in one direction and quick in the other.

The unevenness of the clarinet vibration comes from the fact that the reed is completely closed for a small time in each cycle. During the times when the reed is closed, the energy being given to the column of air in the tube of the clarinet is turned off. We don't hear these "off" moments because they occur for minuscule periods of time, hundreds of times a second. Nevertheless, the regular interruption of the power supply to the instrument means that the pressure ripples it produces will be far more complicated than just a regular up and down pattern—and we hear this as a rich or complex timbre.

However, don't let all this chit-chat about "unevenness of vibration" give you the impression that clarinets or violins make an inferior sound to instruments which have a smoother timbre, such as the harp or flute. As I said earlier, we find these complicated timbres just as enjoyable as the straightforward ones and often prefer them, because they add an extra layer of interest to the music.

When composers are writing something for an orchestra to play, they have to bear in mind the timbres and loudnesses of the instruments at their disposal and then distribute the musical jobs accordingly. This process, called *orchestration*, can make dull music interesting, or interesting music dull, depending on how well it is done. The books on the subject will tell you, for example, that the range of a bassoon can be divided into three parts. Its low notes sound full and rough, its middle notes sound mournful and the upper notes are pale and soft. Other bits of advice include such snippets as the fact that the clarinet can play more quietly than a flute and that a triangle can't be played quietly at all. Most of these books are, of course, measured and scholarly in tone, but my favorite one is almost rabid in its opinions. Professor Frederick Corder wrote his *The Orchestra, and How to Write for It* in 1894. Let's hear his opinion of the trumpet:

I desire here to record my emphatic opinion that the trumpet in the orchestra is an almost unmitigated nuisance. In the small orchestra of Haydn and Mozart it obliterates everything else, and dare only be used here and there in the padding; in the modern orchestra it is useless because of its limited scale, while in the music of Bach and Handel it is a source of constant vexation of spirit.

Sorry Frederick, I wouldn't want to vex your spirit. I'll put the trumpet away. Perhaps a little guitar music?

The guitar is not worth wasting words over, as its very weak tone and deep pitch...

Oops! Perhaps a relaxing melody on the viola?

Viola players have always been both scarce and bad.

Oboe?

The tone of the oboe is thin, penetrating and exceedingly nasal. It is plaintive and pathetic or quaint and rustic according to the character of the music, but should not be heard for too long together.

Pub? Yes? Hang on, I'll get your coat...

One of the few instruments Professor Grumpy has a good word for is the clarinet, but I dread to think what he would have said about the drinking straw oboe. All you need in order to own one of these magnificent instruments is a drinking straw and a pair of scissors. The illustration below shows you what to do. Squash the end of the straw flat, cut it to a point and then put it into your mouth. With the point about one centimeter inside your mouth, use your lips to squash the tube flat while blowing down it. After a couple of minutes' practice you will be able to balance the pressure from your lips closing the tube with the blowing pressure which is trying to reopen it. (If you have a lot of trouble achieving this, you probably have the straw end too far or not far enough into your mouth.) You should get a not-very-mellifluous reed instrument noise. If you cut the length of the tube, you will get different notes as the resonating length gets shorter. You can even cut little finger holes and play dreadful out-of-tune melodies. The long winter evenings will just fly by.

A drinking straw oboe. Squash one end of a drinking straw flat and then cut it to a point. Put the pointed end in your mouth and keep it squashed flat with your lips while you blow into it. Your lips should be positioned approximately where the "squashed flat" arrow is in this sketch. (Paper straws work better than plastic ones because they flatten more easily.)

Tuned percussion

The glockenspiel

The glockenspiel is a member of the "tuned percussion" group of instruments: "percussion" because you hit them to make a noise and "tuned" because they produce notes rather than the untuned bangs of most other percussion instruments such as bass drums. The word glockenspiel means "bell play" in German—but the instrument looks rather like a keyboard made of metal bars on supports.

The way a glockenspiel makes its note is rather simple and is related to how a plucked string moves. When you hit one of the metal bars, you suddenly bend it a little bit and immediately release it. The bar then tries to return to its original straight condition but overshoots and becomes bent in the opposite direction. It then continues to flex to and fro in this way, losing a little energy with each flex, until the note dies away.

From a timbre point of view, the metal bar produces a very pure note—which is almost entirely made up of the fundamental frequency. It manages to do this because of the way it is supported on the glockenspiel. If you take a glockenspiel bar, tie a bit of thread around it and dangle it in the air before you hit it with a stick, you will still get the note, but it will have a more complicated timbre, as the metal flexes in all the different ways it can. If you fit it back onto the glockenspiel you will find that you get the very pure tone again. This is because it is supported at the exact positions which allow the bar to vibrate in only one way—the one which gives the fundamental frequency. This means that all the bending energy can go into the production of this pitch—so it is produced clear and loud. If you moved one of the supports a few millimeters toward, or away from, the middle of the bar, the note would become much quieter and of a more complex timbre. This is because the support would be positioned where the bar needs to move up and down and would interfere with its movement.

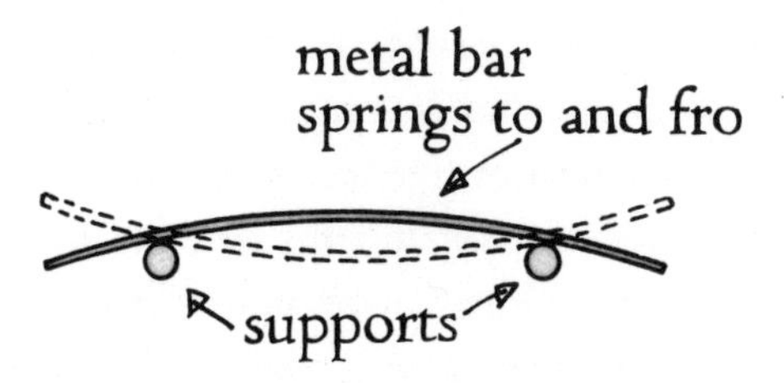

The position of the supports on a glockenspiel bar allows it to flex in only one way—the way which produces the fundamental note. Any other types of flexing are suppressed because they would involve movement at the supports.

The piano

Eleven-year-old boys, and other diligent people who like to collect facts, will delight in telling you that the piano is a percussion instrument. Percussion means that something hits something else to make the sound, but what does that mean when we're describing a piano?

Inside a piano, each key is attached to a set of levers which push on each other and eventually flick a small wooden hammer, wrapped in felt, at one of the strings. The loudness of the note produced is determined by how fast the hammer is traveling when it hits the string. As the hammer approaches the string, it is no longer connected to the levers and this means it can bounce off the string immediately—which it needs to do, otherwise it would rest on the string and stop it vibrating.*

When people talk about a violin player's tone, they mean how

* Actually, each hammer hits two or three strings—all tuned to the same note—for extra loudness, but I will refer to "a string" as if there was only one in each case.

well he produces the notes. This is a combination of several things: the quality of the instrument, how accurately he places his fingers on the neck (and how much he waggles the fingers to get a vibrato effect) and how well he controls the movement of the bow. The violin or flute player has a lot of control over the loudness and timbre of every note he produces from its beginning to its end. This is not true of percussion instruments like the xylophone or the piano. With this type of instrument, you start the note off and let it ring—and you can cut it short if you want to. There is no communication between the note and the pianist while the note is ringing.

So, piano players have a different sort of "touch" from violin players. The instruments are equally difficult to play at a high level, but the skills are different. Pianists can only control how loud a note is, at what time it starts and how long it lasts. On the other hand, she has to control all these things for up to ten notes at a time, which is a hell of an achievement. Highly skilled players can put down all five digits of one hand to make a chord and manage to put down one of those fingers slightly faster than the others in order to make the melody note louder than the others.

It is interesting to realize that quiet piano notes have a different timbre to loud ones, because if you hit a string harder you get a different mix of harmonics. Hitting the string harder tends to encourage the higher numbered harmonics, which gives the notes a more complex, harsher sound. This means that pianists have some timbre control linked to their control of loudness.

The control of the loudness of notes was originally the whole point behind the invention of pianos—and the reason why they are called pianos in the first place. The full name for the piano is the "piano-forte," which means "quiet-loud." There were several keyboard precursors to the piano, but the only one which was loud enough to play with other instruments was the harpsichord.

The harpsichord has a set of keys joined to small spikes made of the quills of crows' feathers, which pluck the strings. The trouble with this system is that it doesn't matter how quickly or slowly you pluck the string, or how much force you use, it makes exactly the same noise at the same volume. On a guitar you can make the plucked note louder by pulling the string further before you let it go—but in a harpsichord the string is always plucked by the same amount so the loudness of the note cannot be changed. This inability to change the volume, and the rather twangy sound of the plucked strings, meant that instrument makers started looking for new ways to excite the strings. Hitting the string with something fairly soft was the winning option and this led to the development of the piano.

The piano was invented in 1709 by an Italian instrument maker with the mellifluous name of Bartolomeo Cristofori, and was continuously developed over the next hundred years or so. Once the instrument makers got the action of the levers worked out, they had an instrument which could play at any volume from quiet to loud. The ability to vary the volume has two great advantages. The first is that you can make the tune stand out from the quieter accompaniment, and the second is that you can vary the loudness (and therefore timbre) whenever you want to, in order to emphasize the emotional climaxes of the piece.

Timbre design—synthesizers

In the 1960s a new breed of musical instrument—the synthesizer—became available to musicians, and rock bands soon found out that their keyboard players were beginning to squander more than their fair share of the instrument budget. Before this point in history, the keyboard player was the one who sat at the back with the drummer and didn't get much post-gig snogging. By the mid-1970s some of them had more dials and switches to play with than the average military helicopter pilot. Synthesizers allowed them to mix harmonics together in previously unheard of combinations to get millions of different timbres. Some of the sounds they produced were marvelous and were clearly the result of weeks of experimentation and planning. Others weren't.

Synthesizers produce musical notes synthetically—that is, there is nothing vibrating inside them; the notes are just produced by combining electrical ripple patterns which drive speakers to produce musical notes. When a natural musical note is produced, the overall ripple pattern is made up of a mixture of harmonics—simple waves join together to produce a complicated wave shape. Electronic engineers use the same principle to produce synthesized notes. Inside a synthesizer the circuitry produces simple ripple patterns which are combined to produce much more complex ones—and because you can choose almost any combination, you have a vast number of timbres to choose from.

Some sounds are more difficult to copy than others with this technology. For example, it is much more difficult to mimic the unmusical sounds which are made by traditional instruments as each note starts than it is to copy the notes themselves. Also, smooth timbre instruments are easier to imitate than complex timbres such as the violin or oboe. Another problem is the fact that if you set your synthesizer to produce a certain ripple pattern, the timbre will remain the same over the whole range of notes from high to low—and, as we saw earlier, real instruments don't do that. Because of these problems, synthesizers are not generally used to mimic other instruments; they are used as instruments in their own right. If you want a traditional instrument sound then you use either a real instrument or sampling technology—which is, effectively, a digital recording of the individual notes from a real instrument.

Something very odd indeed

Look at this collection of frequencies. Together they make up our old friend the note A_2, which has a fundamental frequency of 110Hz:

110Hz, 220Hz, 330Hz, 440Hz, 550Hz, 660Hz, 770Hz, etc.

As you know, the timbre of an instrument is made up of the various loudnesses of these ingredients within the ripple shape. Whatever the mixture of ingredients, our brain recognizes this as a note with an overall frequency of 110Hz. Even if the loudest, strongest component was 330Hz, the overall pattern would only be completing its dance 110 times a second—so the fundamental frequency is 110Hz.

"Yes, John," I can hear you saying, "you've already said all that. Are you being paid by the word or something?" Be patient, dear reader—it's going to get odd in a minute or two.

Rather than just being a minor contributor to the sound, it is possible that one of the harmonics could be completely silent. If, for example, the 770Hz frequency was completely absent, we would still hear the remaining harmonics as part of a note which has a fundamental frequency of 110Hz. This is because only 110Hz can be the head of a family which includes 110Hz, 220Hz, 330Hz, etc. We could have several of the harmonics silent—and still the fundamental frequency would be 110Hz.

Now the odd bit: we can even remove the first harmonic, the fundamental—110Hz—and the fundamental pitch of the note we hear would *still* be 110Hz. This sounds a little insane but it's perfectly true. If you hear the following collection of frequencies: 220Hz, 330Hz, 440Hz, 550Hz, 660Hz, 770Hz, etc. you will hear it as a note with a fundamental frequency of 110Hz, even though the sound does not contain that frequency.

Although the head of the family is absent, the remaining components join together in a dance which repeats only 110 times a second. So the fundamental frequency is 110Hz—and that's that.

A sane person's response is usually that this note should surely sound like the A an octave above 110Hz, the one with a fundamental frequency of 220Hz. But this isn't true because the harmonics of that note are: 220Hz, 440Hz, 660Hz, 880Hz, etc. This group doesn't contain 330Hz, 550Hz or any of the odd-numbered harmonics of the original harmonic family.

These odd-numbered harmonics are present in our group with the missing fundamental—so the only possible "all together now" option for our group is 110Hz.

Not only is this "missing fundamental" thing weird—it's actually useful. Not, perhaps, as useful as a Swiss army knife, or the Heimlich maneuver—but for anything this peculiar to have any usefulness at all is admirable, don't you think?

Hi-fi or even lo-fi speakers have a range of frequencies over which they are effective and this is related to their shape, size and what they are made of. In the old days a good quality speaker cabinet would contain two or three different speakers: little stiff ones for high-pitched notes and big floppy ones for low frequencies. Nowadays it is possible to get ridiculously low frequencies out of small speakers by utilizing the "missing fundamental" idea. Let's say your speaker won't do much at frequencies of less than 90Hz, but you want to hear the note A_1 clearly—and it has a frequency of 55Hz. If you feed the harmonics of 55Hz to your speaker without the fundamental (i.e., 110Hz, 165Hz, 220Hz, 275Hz), you will hear 55Hz loud and clear even though the lowest frequency at which your speaker is moving is 110Hz. Impressive, eh?

You are hearing a note which is not actually being produced.

I told you it was odd.

6. How Loud Is Loud?

Ten times one equals about two . . .

We can all tell when music is getting quieter or louder but it is extremely difficult to say exactly how much louder one sound is compared to another. Trying to decide whether one sound is *exactly* twice as loud as another is just as difficult as trying to decide whether or not you find one joke *exactly* twice as funny as another.

One of the main oddities about loudness has to do with the addition of sounds. Normally when you add things together the result makes sense—if I give Fred one orange and you give him one, then lucky Fred has two oranges; if I give him three and you give him two then he has five. The addition of sounds doesn't work like this. When you listen to a solo violinist playing a concerto with an orchestra, the number of people playing can vary from 1 to 100 in just a second or so, but we don't clamp our hands over our ears and think "Cripes! The music just got a hundred times louder." (The coarser individuals among you may substitute "Cripes" for even stronger language—up to, and including, "Jeepers.") It is difficult to generalize about how much louder the music gets on these occasions: it depends on which instruments are involved and whether or not the composer has asked everyone to play loudly or quietly. It is, for example, possible for the whole orchestra (playing quietly) to make *less* noise than a single instrument playing loudly.

If you add the noise of ten violins (or any other instrument) together, you don't hear ten times the loudness of one instrument on its own. In fact, it's very difficult to estimate exactly how many times louder the sound is, but most people would agree that ten violins (playing the same note with the same amount of effort) sound approximately twice as loud as one. Similarly, for 100 instruments the answer would be, "One hundred times one? . . . That'll be about four . . . Obviously."

So, ten instruments sound only twice as loud as one, and 100 instruments sound only four times as loud as one. These "strange but true" statements need some explanation. Fortunately I have one ready. . . .

Before we start our discussion, let's simplify matters by talking about only one kind of instrument at a time. We'll concentrate on flutes for the next bit, but the following points are also true of any mixed group of instruments.

So, let's imagine we've gathered together an orchestra of 100 flute players. First of all there is silence. Then one of the flutes starts to play a note. The difference between silence and one flute is very impressive—it's rather like sitting in darkness and then lighting a candle. Then another flute starts playing the same note. This makes a difference—but it's not as big as the difference between the silence and the first flute. When the third flute joins in (playing the same note) it only makes a small difference to the volume of the sound and the fourth makes even less difference. If they all play the same note, and the flautists keep on joining in one at a time, you will soon find it impossible to tell when a new one joins the group because the difference between, say, sixty-two flutes and sixty-three is so tiny.

This is all very odd because you could have asked that sixty-third flautist to be the first in the group—and in that case he or she would have been the one who made the biggest difference. In fact, you could ask every one of the flute players to play a solo note as loud as they can after a silence and they would all sound equally loud.

There are two reasons why our 100 flute players sound less loud than you would expect. One of them has to do with how sound waves add together, and the other is related to how our hearing system works. Let's look at these things one at a time.

How sound waves join together

By now, we are all clued up about the fact that a musical note is a regular pattern of changes in air pressure which makes our eardrums flex in and out. The number of times the eardrum flexes every second tells the brain the pitch of the note—and louder notes involve bigger pressure changes, so the eardrum flexes more. (If you overdo it by listening to a very loud noise like an explosion, the high pressure will flex your eardrum too far and tear it, giving you what's called a perforated eardrum.)

The illustration below gives you a picture of what's going on. Both of these ripple patterns have the same frequency—but one of them involves much bigger pressure changes, so it pushes and pulls at the eardrum more and sounds louder.

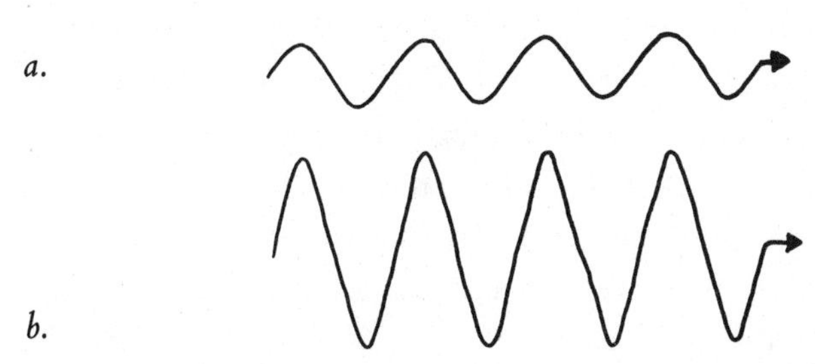

The pressure ripple patterns of the same note: played quietly (top) and loudly (bottom). The frequency of the note has not changed but the variation in pressure is greater for the louder note.

Let's go shopping. You go to the music shop and buy a pair of identical glockenspiels and I'll go off and buy a sound pressure monitor.

This equipment contains a microphone which works just like an ear—sound pressure waves push a part of it in and out like an eardrum—and it has a computer which measures how powerful the waves are. I'm using a sound pressure monitor because it responds in a very straightforward way to changes in sound pressure—if you double the sound pressure you get double the reading from the computer.

Now all we need is a big hotel and a pair of identical twins (trust me, this is eventually going to be interesting). We start in one room and get the first twin to hit any note on the glockenspiel while we measure the power of the pressure waves it produces. Let's say the computer tells us that the loudness of the note just after hitting the glockenspiel is ten pressure units.

Now we take the second twin to any other room in the hotel and get him to hit the same note on the other glockenspiel with the same force as his brother. (We picked identical twins because we want them to hit with the same force as each other.) As you would expect, when we measure the pressure ripples the computer says that the loudness is, once again, ten.

Now we get them both together in the same room. First of all they take turns hitting their notes—and, not surprisingly, there is no difference in the reading we get; as long as they hit the same note with the same force, we always get a reading of ten.

Finally we get them both to hit the note at the same time. We might now expect the computer to say that ten plus ten is twenty. But it doesn't. We can do this a few times and the average sound pressure measured for the combination of the two notes will be about fourteen.... Some of our sound has gone missing.

And if we went out and bought more glockenspiels and hired more twins, we would find that, for forty instruments, instead of getting a pressure reading of 400, the result would only be sixty-three!

Widespread disappointment—we have a roomful of expensive

twins and glockenspiels but a lot of the sound is simply disappearing. Let's send them all down to the hotel lobby for afternoon tea while I explain what is happening.

When we hit only one instrument we get the best value for our efforts: we hit the bar, it wiggles up and down, and passes on these wiggles to the air as ripples in air pressure. So we get our money's worth from the hit.

If we have two instruments, we only get double the effect if the up–down–up–down pressure ripples from them are perfectly in step with each other—so they can act together to give an UP–DOWN–UP–DOWN pressure ripple.

But, when we hit both instruments, you can bet your life that we don't hit them *exactly* at the same time, so the pressure ripples from the two instruments won't be in step when they reach the microphone. This means that sometimes the "pressure up" part of one ripple will be trying to raise the air pressure as the "pressure down" part of the other is trying to lower it. If the wave patterns were perfectly out of step, the up–down–up–down of one of them would be canceled out by the down–up–down–up of the other—and we wouldn't hear a note at all.

This is weird but true—it's how some farmers protect their hearing when they are driving noisy tractors all day. They buy "active ear defenders" which look like headphones. Inside each of the earpieces is a microphone and a speaker connected to some electronics. The microphone listens to the sound which is about to reach your eardrum and makes the speaker produce the same pressure wave—but out of step with the original one. The idea is that when the two pressure waves meet, one of them tries to raise the pressure at the same time as the other tries to lower it—so nothing much happens and the eardrum is left in peace. In practice the sound waves are too complicated for this to work exactly, but it does reduce most of the noise.

Going back to our glockenspiels, the canceling out is nowhere near perfect because it would be too difficult to organize—the sound waves are coming from different places in the room and also bouncing off the walls, and it's incredibly unlikely that you would hit the instruments at precisely the right times to get the ripple patterns exactly out of step just at the point where they meet the microphone. What actually happens is that we do get more sound pressure from two instruments than we would from one—but there is some interference from the low-pressure bits of one wave pattern with the high-pressure bits of the other, so there is some canceling out.

If more instruments are involved, the amount of canceling out gets more serious. The pressure of the air next to the microphone can only be higher than normal (pushing the microphone inward) or lower than normal (pulling it outward): it can't be both at once. Each of our forty glockenspiels has an "up pressure" or "down pressure" vote at any point in time—but a lot of these votes cancel each other out. If a forty-first glockenspielist joins our little party, then his note will be mostly canceled—though a little bit will get through to contribute to the overall loudness.

This effect is not the only one involved in our appreciation of loudness. If it was, 100 instruments would sound ten times as loud as one. But, as I said earlier, we perceive 100 instruments as being only four times as loud as one. This extra diminution in perceived loudness is the result of the way we humans are designed—so let's have a look at that.

Why our brains don't add up sounds properly

Why don't our brains add up sounds normally? The surprising answer is that our brains and ears add up sounds in an unusual way in order to help us stay alive. From the times of the earliest cavemen to the present day, we have used our ears to help us avoid danger. This is one of the main reasons we have ears in the first place (although they are also useful for supporting your sunglasses). To be effective, your ears have to be able to hear very quiet noises (like the sound of someone creeping up on you), but also they must not get damaged by loud noises (such as thunder). It wouldn't be any good if you had excellent hearing for quiet noises but your ears stopped working after the first loud noise you heard.

Our ears are organized in such a way that quiet noises can be heard clearly but any increase in the volume of the noise has progressively less and less impact. This effect is also true of our other four senses: smell, taste, sight and touch. Six smelly socks aren't six times as smelly as one on its own (even though each of the socks is releasing the same amount of smell) and ten salted peanuts in your mouth aren't five times as salty as two of them (even though you now have five times as much salt on your tongue). If you light 100 candles one at a time in a dark room you get the same effect as you got with the flutes—the first one makes the biggest difference and the eighty-seventh makes hardly any difference. If you are daft enough to stick a pin in your fingertip then it will hurt, but if you stick a second one in (next to the first one) the pain will not be doubled.

Why, you may ask yourself, did I carefully point out in that last sentence that the pins should be next to each other? Well, there is a reason, and it is surprisingly relevant to our discussion about the loudness of sound. Imagine that I have accidentally trodden on a thumbtack with my big toe. Obviously I would feel quite a lot of pain and would probably do a fair bit of swearing about it. If I trod on two thumbtacks with my big toe, the overall sensation would only be a little worse than a single tack—nothing like twice as bad. If, on the other hand, I trod on one thumbtack with my left big toe and another with my right big toe, then the pain would feel much worse than two tacks in one toe (don't try this at home—just take my word for it). The reason for the increase in pain is the fact that my brain would receive two distinct pain signals (one from each foot) rather than a "two-tack" pain signal from one toe.

What has all this got to do with music? Well... earlier on I said that the ear/brain would calculate ten flutes as being only about twice as loud as one flute. This is only true if the ten flutes are playing the same note. If you divide the flutes into two groups and ask group 1 to play a note with a much higher (or lower) pitch than group 2, then the two notes played together sound louder than when everyone is playing the same note. The difference in pitch between the two notes needs to be bigger than the difference between "Baa" and "Black" for this effect to work. One reason for this apparent increase in volume is the fact that the brain is now receiving two distinct sound signals (like the two pain signals). The other reason is that the notes from the two smaller groups experience less of the "canceling out" effect I mentioned earlier.

Loudness and pitch

The sensitivity of our hearing system is not the same at all frequencies. The most extreme demonstration of this is the fact that there are some sounds that we can't hear at all because they have a pitch which is too high (e.g., the note from a dog whistle) or too low (e.g., the subsonics you sometimes experience if a large truck engine vibrates the windows of the building you are in). Both the dog whistle and the trembling windows produce a note or noise—it's just that our ears are not designed to hear them. Even within the range which can be heard by the human ear there are differences in sensitivity. We are most sensitive at the rather high, squeaky frequency range covered by the top few notes on a piccolo—which is why you can hear a piccolo clearly above the other instruments of an orchestra or marching band. In fact, music textbooks advise composers that the piccolo should be used only sparingly because it is difficult to blend it in with the other instruments.

At frequencies higher or lower than this high, squeaky range, our ears are progressively less sensitive. Most musical notes are below this range. This means that if you want to get a balanced sound between a bass instrument like a bassoon and a higher pitched instrument such as a clarinet, the bassoonist might have to play as loud as he can while the clarinetist takes it easy. Similarly, if two identical instruments are playing together but one is playing high notes and the other is playing low notes, then the one playing the low notes must play harder in order to sound as loud as the one playing the high notes.

Loudness and note duration

Yet another peculiarity of loudness is that of note duration. The normal loudness of the note can be heard if it is played for a second or so, but if the note is played for half a second or less, it will sound quieter. (It should be borne in mind that lots of music involves notes which are shorter than half a second; for example, when we sing "have you any wool?" only the word "wool" is longer than half a second.)

On the other hand, if a note is played for several tens of seconds then its loudness will appear to decrease as the brain begins to stop noticing it so much. This effect, of diminishing intensity for a continuous stimulus, also happens with our other senses, particularly our sense of smell (which is something we can be quite glad of at times). The reason why the sound appears to diminish after a while is that your brain is constantly monitoring your senses for danger signals. If a sound is continuous, and nothing bad is happening, your brain loses interest because the noise is obviously not important to your well-being. Your brain is primarily interested in any sudden changes in the sounds you are hearing, which is why you sit up and take notice if a long-lasting sound suddenly stops— the "deafening silence" effect.

Measuring loudness

Human beings like measuring things—we measure our height, weight, the speed of our cars and the size of our bathrooms. Measurements help us to discuss things more accurately and clearly. There are, of course, a lot of things we can't apply accurate measurement systems to, such as kissing ability, or the social skills of hamsters, but whenever we can we invent and use a measurement system. As we shall see, the invention of a measurement system for loudness was nearly as tricky as getting one to work for kissing (and probably a lot less fun). Before we start this section I would like to go over a couple of points about measurement systems in general.

There are two basic types of measurement systems: the absolute type and the comparative (or relative) type. If we are using the absolute type, we would say "Farmer Smith has eight cows and Farmer Jones has four cows." If we use a comparative system, we would say "Farmer Smith has twice as many cows as Farmer Jones." As you can see, both systems give us some useful information, but the absolute system is more precise, which is why we normally use it. In some cases, however, we can't use the absolute system and we need to use the comparative one. Yes . . . you guessed it . . . loudness is one of these awkward cases.

Because our ears respond to pressure changes, any system for measuring loudness should be based on the measurement of pressure. Unfortunately, however, the earliest system of loudness measurement was adapted from a method for measuring the decrease in strength of electrical signals after they had traveled down a mile of phone cable, so we ended up with a system based on intensity rather than pressure. This is rather like measuring distances in gallons of gas (if New York to Boston uses fourteen gallons of gas, then the distance from here to Poughkeepsie is six gallons). The numbers are useful and accurate in their way, but it's all a bit clunky. This energy-intensity system of measuring the loudness of sounds has advantages and disadvantages, as we shall see.

Measuring the intensity of sounds

Remember our twins with their trusty glockenspiels? They would get just as tired hitting the instruments in individual rooms as they would if they were standing next to each other. In each case they are using the same amount of energy and it's not their fault if the pressure waves refuse to cooperate fully. The intensity system looks at how much energy they both put into their bonging, rather than at how much sound they make. This system says to itself: "The bonging energy involved doesn't change just because they both now share a room—one bong plus one bong equals two bongs' worth of energy intensity." This convenient ability to use simple addition is the main advantage of the intensity system of loudness measurement.

If we take a microphone, attach it to a computer and ask it to convert the pressure readings it hears into energy-intensity measurements, we can add sounds together by the normal rules of addition. A computer can be programed to recognize that the ten flute players are working equally hard to produce ten times as much sound intensity as a single flute. So, for example, we could now say: ten violins produce ten times the sound intensity and twice the loudness of one violin.

Let's say we are going to use a computer and microphone to measure sound intensity from total silence to painfully ear-damaging.

After some careful experiments we could find the quietest noise that a human can hear. We could then set the computer so that it gave this sound intensity a value of "1" and call it the threshold of hearing. This noise would be, perhaps, equivalent to someone ten yards away sighing. If ten people were sighing ten yards away (let's not go into why they are all so unhappy), then the computer would give this new sound an intensity level of 10 (but, of course, we would hear it as only twice as loud).

Moving up in loudness, we can abandon all those miserable sods and start measuring the sound intensities of motorbikes or brass bands. You might imagine that, by the time we get to the noise levels which cause pain (e.g., putting your ear a few centimeters from a road drill), we would be measuring sound intensities a few hundred times greater than our original sigh. Well, stand by to be flabbergasted—the sound intensity which causes pain is 1,000,000,000,000 times greater than that of the quietest noise you can hear—yes, the sound intensity generated by a road drill is a trillion times greater than that of a sigh. So if you're unhappy with your job as a road drill operator, and you want your sighs to be noticed, remember to turn the drill off first.

We need a quick reality check here. As I said earlier, our ears do not directly measure intensity—they monitor pressure differences. Pressure differences are related to intensities, but to convert the intensity to pressure we need to do a calculation. In this case the calculation tells us that the range of sound pressure difference between near silence and pain is not 1,000,000,000,000—it's just 1,000,000, a million. It's still an enormous number but it's not a ridiculously enormous number.

Let's re-enter the world of intensity measurement... We know that every time we multiply the sound intensity by ten (by having ten violinists play rather than one), then the loudness of the sound doubles. So—let's put together a list of sounds starting from the quietest to the loudest we can hear. Every sound in this list is twice as loud as the one above it.

A list of sounds from the threshold of hearing to the threshold of pain

Example	Relative loudness	Relative sound intensity
Almost silence (a sigh 10 yards away)	1	1
A small fly in the room	2	10
A large bee in the room	4	100
Someone nearby humming a tune	8	1,000
A fairly quiet conversation	16	10,000
Solo violin, moderate volume	32	100,000
A busy restaurant (or ten violins)	64	1,000,000
City traffic, rush hour	128	10,000,000
An orchestra playing loudly	256	100,000,000
Very noisy nightclub	512	1,000,000,000
Close to the speakers at a rock concert	1024	10,000,000,000
Big fireworks explosion	2048	100,000,000,000
Pain—a few inches from a road drill	4096	1,000,000,000,000

(All my examples are, of course, just guidelines—perhaps you have raucous, noisy bees where you live—or maybe your sister is a spectacularly loud hummer.)

This table illustrates some of the main points about loudness and gives us two methods for comparing loud and quiet noises. However, neither system provides us with a useful numerical scale of loudness because the numbers are so large. The relative sound-intensity numbers on the right say that a fly has a value of 10 and the violin has one of 100,000, which means that you would need 10,000 small flies in the room to produce the same sound intensity as a violin. This is useful stuff if you are a violin-playing maggot farmer, but we still need a noise level measurement system which uses a smaller range of numbers.

Decibel madness

The search for a system with a small range of numbers gave someone, sometime in the first half of the twentieth century, the clever idea of a loudness measurement scale based on how many zeros there were after the "1" in the "relative sound intensity" column in the table above. This is the "Bel" scale and in it a sound intensity of 1,000 would have a loudness of 3 Bels and 1,000,000 would be 6 Bels, etc. (just count the zeros in each case). This was thought to be a brilliant idea for about seven and a half minutes until someone even cleverer pointed out that this would only give us twelve numbers of loudness measurement from extremely quiet to painfully loud and that this wasn't going to be a very useful system—because now we didn't have *enough* numbers.

Finally it was decided that 120 measures of loudness would be more useful and that could be achieved by multiplying all the numbers in the Bel system by ten and thus measuring loudness in "tenths of a Bel" or "deci-Bels" (*decibels*). So now we have a system where a sound intensity of 1,000 is equal to 30 decibels and 1,000,000 is 60 decibels, etc. (count the number of zeros and multiply by ten). A rough sketch of what the decibel system means to our ears is presented in the table below ("decibel" is usually abbreviated to "dB").

Intensity, decibels and loudness

Relative sound intensity	Decibels (dB)	Relative loudness
1	0	1 (almost silence—sigh)
10	10	2 (small fly)
100	20	4 (large bee)
1,000	30	8 (humming)
10,000	40	16 (quiet conversation)
100,000	50	32 (solo violin)
1,000,000	60	64 (busy restaurant)
10,000,000	70	128 (rush hour)
100,000,000	80	256 (loud orchestra)
1,000,000,000	90	512 (nightclub)
10,000,000,000	100	1024 (rock concert speakers)
100,000,000,000	110	2048 (big fireworks)
1,000,000,000,000	120	4096 (pain—road drill)

Now we have a system that can be used to measure loudness from the quietest noise to the loudest which only goes from zero to 120. But I'm afraid that even though the numbers are now simple, the use of this scale is complicated. The table shows that each time the loudness of the noise doubles, you add 10 decibels. This sounds simple enough until you realize that this means that not only is 20dB twice as loud as 10dB (which seems obvious) but also that 90dB is twice as loud as 80dB (which seems crazy, but it's true—just look at the table).

At this point I must come clean and admit that I don't like the decibel system of loudness measurement at all. It isn't easy to use even if you have studied math or physics up to college level. Even a professional scientist would need a calculator and a few minutes to be able to tell you the difference in loudness between 53 decibels and 87 decibels. I have no proof of this, but I think the decibel was invented in a bar, late one night, by a committee of drunken electrical engineers who wanted to take revenge on the world for their total lack of dancing partners. Apart from the calculation problems, the use of intensity for measuring sounds is indirect and overly complicated.

I started this book with the promise that there would be no mathematical formulas and I intend to keep that promise. However, I can't explain how to calculate the difference between 53 and 87 decibels without using formulas. Anyone who would like a bit more information on using the decibel system will find some in part B of the Fiddly Details section at the end of this book. Frankly, though, I think we should leave the horrid decibels behind us, and move on to discuss the much more user-friendly systems developed in the 1930s by a bunch of American researchers, led by an experimental psychologist called Stanley Smith Stevens.

Better loudness measuring systems: the phon and the sone

While the electrical engineers were giggling to themselves about lumbering us all with the decibel system, the sound engineers, concert hall designers and psychologists specializing in hearing decided to strike back. Because they were dealing with loudness measurements all day, these people knew that the decibel system had two big flaws:

Big flaw 1. As I said earlier, the human hearing system is more sensitive at some frequencies than others. This means that a 32dB high note from a flute will sound (to a human) a lot louder than a 32dB low note from a bass guitar—so the decibel system is an unreliable measure of loudness for human beings.

Big flaw 2. Before the advent of pocket calculators, you had to sit up all night with six pencils and three erasers working out stuff like, "How much louder than 49dB is 83dB?" Even after calculators came along you had to buy one with a farcical number of buttons on it and an instruction book as big as Webster's Dictionary.

"Aha!" said the experimental psychologists, "we can eschew the decibel system and develop one which gives a more accurate picture of the response of the human ear." (Experimental psychologists talk in this superior, pedantic way whenever they get the chance.)

The only way to develop a system based on the subjective response of the human ear is to carry out tests on lots of people. This is why the work was carried out by psychologists: they were measuring people's opinions, rather than things which can be measured by scientific equipment.

The first set of tests was designed to overcome big flaw 1 and involved a large number of people who were asked to compare notes of different frequencies and say when they thought they were equally loud. These tests were carried out over a wide range of loudnesses and frequencies and led to the development of a unit of loudness called the *phon*.

To explain the difference between phons and decibels, let's imagine that we have enlisted a robot to play the piano for us. First of all, we ask the robot to produce a loudness of 50 decibels for each note and play all the notes, one at a time, starting from the top one.

Because our ears are progressively less sensitive as we move from the high notes on a piano to the lower ones, we would hear the notes getting gradually quieter as the robot moved down the keyboard. On the other hand, a computer attached to a microphone, "listening" to the same notes, would "hear" each one as having the same energy intensity.

Now we ask the robot to do the same thing again, but this time producing a loudness of 50 phons from each note. This time our mechanical pal would hit the notes more and more forcefully as it moved down the keyboard. You would hear all the notes as being equally loud because the robot would be compensating for the fact that you can't hear the low notes on a piano as easily as the high notes (your computer would "hear" exactly what the robot is doing and would be able to tell that it was playing the low notes louder).

The phon system is just the decibel system after we have compensated for the fact that human ears are less sensitive to low notes.

Having conquered Big flaw 1 the psychologists moved on to Big flaw 2. As the phon system is basically a version of the decibel system, we still have exactly the same problem working out what the numbers mean. What's the loudness difference between 55 and 19 phons? Pass me a calculator and some chocolate chip cookies...

So the psychologists decided to drop the decibel system entirely and find a way of using a scale based on relative loudness, much to the annoyance of the electrical engineers who developed the decibel system. No party invitations have passed between the two groups since 1936.

If you look back at the table on page 95, then you will see that the relative loudness scale goes from "1" (a sigh) to "4096" (a road drill) and, as we saw, this involves too many big numbers to be a useful scale. But hang on a minute, the numbers for loud noises are only this big because we started with 1 as the quietest noise we can hear. This is the same as measuring the price of everything in pennies—we don't say "my car cost 1,500,000 pennies." We don't need to use the smallest possible coin as the basis for our measurement system. The psychologists gave this some thought and decided to move the "1" from an extremely quiet sound to somewhere nearer the middle of our hearing range. They moved it to the level of a fairly quiet conversation—so now we have a new measurement range with "1" where "16" used to be. This means we have to divide all the numbers in our "relative loudness" column by 16 and the numbers now only go up to 256, as you can see in the next table. Although we have to use fractions of 1 for quiet noises, this is not much of a problem because we don't need to discuss quiet noises very often.

So now we have a system which really works for humans—it's called the *sone* system, and there is no need to worry about big numbers or complicated math. Eight sones sounds twice as loud as 4 sones and 5 sones sounds half as loud as 10 sones.

Modern loudness meters *should* measure different loudness levels in sones because it's the most sensible system for monitoring and discussing loudness levels as far as humans are concerned. For example, if an acoustic guitar player has a loudness level of 4 sones and a rock band's level is 40 sones, it means that the rock band is ten times louder for any human listener. Because the sone system is based on human hearing, sone-based loudness meters automatically make the necessary compensations related to the frequencies of the sounds they are monitoring. Meters like this are used to help the relevant engineers to develop better loudspeakers and sound insulation materials.

However, you may have noticed my use of the word "should" at the beginning of the last paragraph. In fact, most of today's loudness measurements are carried out using the decibel system, although there is usually an adjustment for the human sensitivity to different frequencies. This is simply because the decibel system was the first to become established and is the one referred to in the government documents and legislation dealing with noise levels and soundproofing. So we're stuck with it.

The sone system of loudness measurement

Example	**Relative loudness (after compensating for frequency)**	**Sones**
Almost silence (a sigh)	1	0.06
A small fly in the room	2	0.12
A large bee in the room	4	0.25
Someone nearby humming a tune	8	0.5
A fairly quiet conversation	16	1.0
Solo violin, moderate volume	32	2.0
A busy restaurant (or ten violins)	64	4.0
City traffic, rush hour	128	8.0
An orchestra playing loudly	256	16.0
Very noisy nightclub	512	32.0
Close to the speakers at a rock concert	1024	64.0
Big fireworks explosion	2048	128.0
Pain—a road drill	4096	256.0

But one day we will all rise up and, impaling our scientific calculators on especially pointy road drills, we will free ourselves from the evil, oppressive shackles of the ridiculous, loathso—

But perhaps my cold, authorial objectivity is slipping a little here. . . . Let's move on to a subject less troubled by controversy: the art and science of harmony.

7. Harmony and Cacophony

Tuneful babies

Babies sing little songs to amuse themselves, which often consist of one note repeated over and over again. As they get older they increase the number of notes involved because one-note songs are a little uninspiring. As babies get more adventurous, they will discover the lowest and highest notes they can sing, and find that they can produce any note at all within this range.

The baby, singing its "La La La" song, will choose any old notes within its range and may wander from pitch to pitch, without hitting the same note twice. A "song" of this type might involve hundreds of slightly different notes—and so it can never be repeated. As the child gets older she will realize that everyone else is singing songs which can be memorized and repeated because they involve a limited number of notes: "Baa Baa Black Sheep," for example. The child will hear quite a few people singing "Baa Baa Black Sheep" and will eventually realize that it isn't important what note you start on—it's the jumps up and down in the tune which matter. To make a song recognizable all you have to do is sing the correct size jumps with the appropriate rhythm.

Once a child remembers a few tunes in this way she will develop a library of tuneful jumps that she can use for any song. For example, the jump between "Baa" and "Black" is the same as the one between the "twinkles" in "Twinkle, Twinkle Little Star." The jump between "have" and "you" (as in "have you any wool?") is the same as the one between the first two notes of "Frère Jacques."

The child is now using scales, that is, a limited number of recognizable jumps in pitch. As I said earlier, these jumps are called *intervals*. Trained singers and people with natural ability manage to sing these intervals accurately but the rest of us just get close enough for the melody to be recognizable.

The simplest type of music involves a single voice singing a series of intervals, one after the other, to produce a melody. The next obvious step is to get several friends around to sing along with you—with everyone singing the same notes. This sort of music-making has been around since we were all living in caves waiting for someone to invent central heating.

The earliest cavemen who sang together were quickly followed by the second earliest cavemen who sang together—who decided to liven things up a bit. Like all teenagers they wanted to have their own style of music and didn't care for the old-fashioned rubbish their parents were singing. They had a lot of success with a new technique where half of the tribe sang the song while the other half sang just one note, called a *drone*. They noticed that some of the notes of the song sounded better with the drone than others, but they didn't give it much thought. Eventually, a particularly talented cavewoman called "Ningy, the particularly talented singer" started to sing along using different notes from everyone else. This meant that you could hear two tunes and a drone at the same time. Everyone was delighted, and people became less fretful about the total absence of weatherproofing.

Ningy the particularly talented singer knew she had to pick her notes carefully if they were to sound good with the notes sung by everyone else—some combinations sounded great but others were terrible. This careful choice of notes which sound good together gives us *chords*, and chords are the basis of *harmony*. When I say "notes which sound good together" I don't just mean nice, pleasant combinations. As we shall see, harmonies are not always harmonious and it is a composer's job to build up tension occasionally and then relax it. The American rock musician Frank Zappa summed this up excellently when he said that music without an ebb and flow of tension would be like "watching a film with only good guys in it."

What are chords and harmonies?

Chord: A chord is the sound made by three or more notes played at the same time.

Harmony: A succession of chords produces a harmony. The relationship between chords and harmony is therefore similar to that between words and sentences.

When composers (and by composers I mean people who write pop songs and advertising jingles, as well as Mozart and Co.) write a piece of music, they usually use harmony to provide a background to the melody. This harmony can alter the mood of the melody just as the background of a photograph can make a portrait more or less cheerful. Film music composers often use only three or four tunes for an entire film, and they need to change the feel of the melody to match the moods of the different scenes. Techniques for altering the mood of the music include using different instruments (if it's a scene in Paris we will hear the obligatory accordion) and playing the melody faster or slower. But playing the tune with a different harmony is one of the most effective ways of manipulating our emotions.

Some combinations of notes sound pleasant and some sound tense or ugly. Composers often deliberately choose a sequence of anxious-sounding chords to build up tension before releasing it with some harmonious combinations—composing is rather like telling a story or joke, in that the composer needs to set up a situation and then resolve it in some way. Changing the level of tension in the harmony is one of the composer's main tools for manipulating the mood of the music. An excellent example of this can be found in the opening to a track by Genesis called "Watcher of the Skies," which begins with a series of slow chords played by the keyboard player. In the first part of this solo there is no tune or rhythm to speak of, but tension is cleverly built up by the use of harmony alone. If you care to count the chords, you will find that the thirteenth one introduces us to a whole new level of tension—perfect for accompanying the sort of teenage angst I was attempting to experience when this recording came out.

If you are in the mood for some *really* anxious chords, try listening to "The Devil's Staircase," a piano piece by the composer György Ligeti. It only lasts about five minutes, but by the end of it you don't know whether you want to have a quiet lie-down in a darkened room or listen to it again. I usually can't resist another go before I start looking for a soothing lava lamp.

Inharmonious chords can also be used for comic effect—just listen to the piano at the beginning of the song "Driving in My Car" by Madness. In the few seconds after the first four horn beeps, and before the singing starts, the pianist chooses a lot of jangly chords to add to the general chaos.

As for pleasant chords, there are so many examples that it's difficult to know where to start. Harmonious chords dominate the musical landscape, from "Miss Chatelaine" by k. d. lang to "Adagio for Strings" by Samuel Barber.

But the subject of harmony can be boiled down to a single question: why do certain notes sound good together?

The illustration opposite shows the pressure waves of a note reaching the ear. This drawing is a great simplification because it only shows the fundamental frequency of the note and, as I described in chapter 3, a real note would be a lot more complicated. But I will use these simplified drawings to keep things as clear as possible.

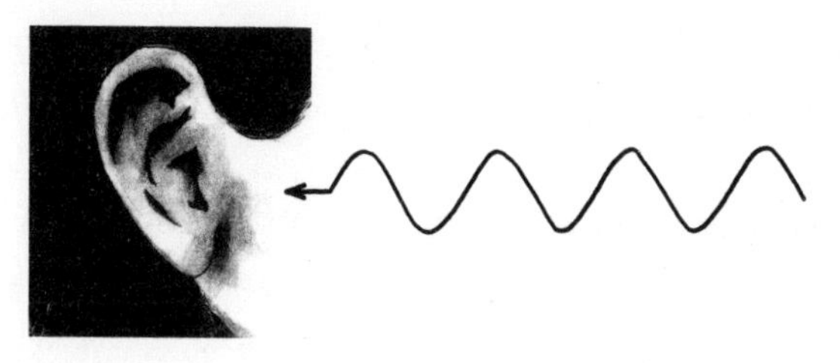

A note (ripples of air pressure) traveling toward an ear. The ripples will make the eardrum vibrate backward and forward like a mini-trampoline. The vibration repeats as a regular pattern and will therefore be experienced as a note by the brain.

The illustration above shows us the fundamental wave pattern of a single note, but if two notes are played at the same time they join together to make a combined wave pattern like the one in the drawing below. If the two waves join together in a regular, orderly way, the combined sound will be smooth and harmonious, and we can see that this is what has happened here.

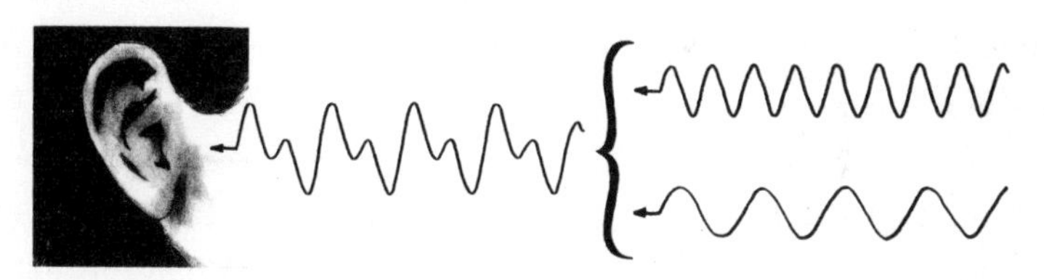

The combination ripple pattern of two notes one octave apart: it looks good and sounds good because the relationship between the note frequencies is very simple—one note has twice the frequency of the other.

In this case the higher note has a frequency which is exactly twice that of the lower note—and the interval between such notes is called an *octave*. If you hear these two notes at the same time, they sound so comfortable together that they are difficult to distinguish. In fact, music psychologists in spotless white lab coats have decided that two notes an octave apart are so closely related that they are almost identical as far as the human brain is concerned.

Let's look at an example of how this works. The third G from the left on a piano is called G_3. If you play this note in front of a choir and ask them to sing it, a lot of the men with low voices will sing the note an octave below yours (G_2), and a lot of the women with high voices will sing the note an octave above (G_4). If you tell them that they were singing the wrong notes they will reply (huffily) that they were not: you asked for a G and that's what they gave you.

The reason why notes an octave apart sound so similar is easy to understand if we refer back to the nature of musical notes. In chapter 3 I explained that a musical note is made up of a family of vibrations: a fundamental frequency together with twice, three times, four times, five times, etc., that frequency.

So let's look at the frequencies we hear if we play the note A_2 (110Hz) and the note an octave above it, A_3 (220Hz).

A_2 (110Hz) is:	110	220	330	440	550	660	770	880Hz, etc.
A_3 (220Hz) is:		220		440		660		880Hz, etc.

So if we hear the 110Hz note first and then both of them together, the brain is not provided with any new frequencies—it just gets a double dose of some of the frequencies it heard in the original note. For this reason the frequency-recognizing system of the brain hears the combined notes as simply a slightly different version of the first note—which is why notes an octave apart sound so harmonious.

This very strong family relationship is the reason why notes an octave apart are even given the same name (well, almost exactly the same name: as you know, we give them a letter with a number relating to where they are on the piano keyboard).

Since notes an octave apart agree with each other completely, the result is so harmonious that it might even be considered a little uninteresting. There are, however, other combinations which sound good without the notes involved swallowing each other's personalities. We can illustrate this by using the notes of "Baa Baa Black Sheep."

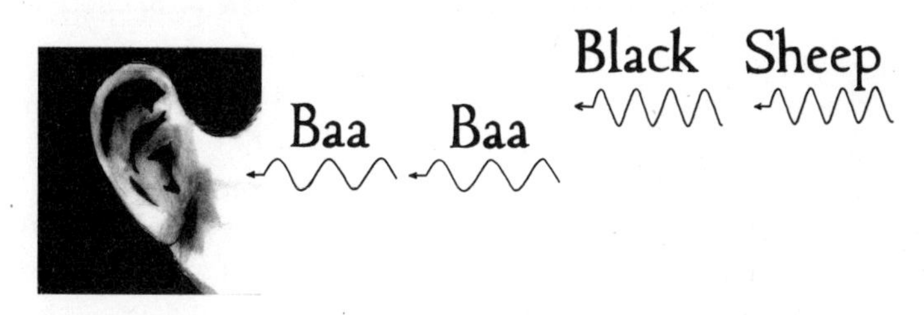

The notes "Baa Baa Black Sheep" traveling toward an ear.

Here you can see that the first two notes are the same as each other and the second two notes are also a pair—but they are different from the first pair. The peaks of the ripples of the second two notes are closer together than those of the first pair and these higher frequency ripples make the eardrum move backward and forward more frequently (which is why we hear a higher pitch note).

When we listen to one person singing this song we obviously hear the notes one after the other, but if we get two singers and ask one of them to sing the note for "Baa" and the other to sing the note for "Black" at the same time, they will sound very good together. In the case of these notes, the "Black" has a ripple frequency which is one and a half times that of the "Baa." Because of this simple relationship, these two notes sound almost as pleasant together as two notes an octave apart do. The next illustration demonstrates how the two notes combine to make a new wave pattern which is smoothly repeating enough to sound good, but different enough from the original notes to sound interesting.

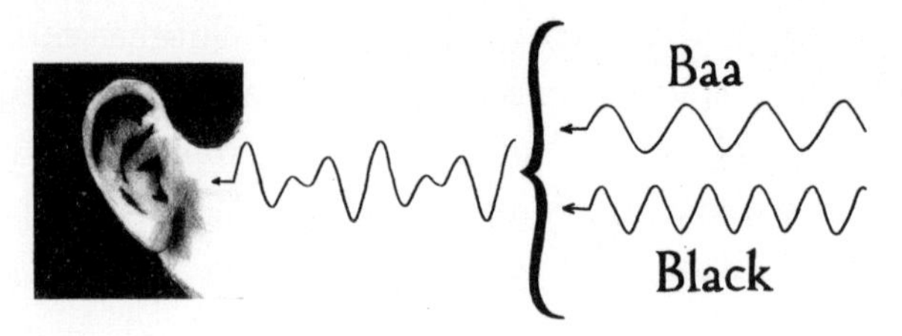

The combination of two notes moving toward an ear. The ripple patterns of the two individual notes join together to become a combined pattern. In this case, the note for "Black" is one and a half times the frequency of the note for "Baa." This simple relationship makes the combination of the notes sound smooth and pleasant.

So far we have looked at combinations of notes which have a simple relationship, such as double or one-and-a-half times the lower frequency. Now let's see what happens if we combine the ripple patterns of two notes which have a complicated relationship to each other. In the illustration opposite you can see the combined ripple pattern created by joining two notes together where the fundamental frequency of the lower note is seventeen-eighteenths the frequency of the upper one. Two adjacent notes on a piano have this relationship and, if you play any such pair together, it does sound jangly and inharmonious. Adjacent notes on a piano are only a semitone apart in pitch, which is the smallest interval we use in Western music. If you play two notes this close together at the same time, the resulting combination sounds as if the two notes are simply out-of-tune versions of each other—competing for our attention rather than supporting and adding interest to each other.

One effect of playing two notes which are close together in pitch is that the loudness of the combined sound goes up and down several times a second. You can see this in the illustration opposite, as the overall size of the combined pressure ripple continuously swells and shrinks. This effect is caused by the individual pressure ripple patterns from the two notes repeatedly falling in and out of step with each other. To understand how this happens, imagine you are walking next to a friend who takes slightly longer strides, but he takes slightly fewer strides per minute, so you are walking at the same speed. If you start off in step with each other, then you will gradually fall out of step, but after a certain amount of time your steps will synchronize again. This will happen in a repeating cycle—maybe he takes ten steps in each cycle and you take eleven. When this "in step–out of step–in step–out of step" cycle happens with the pressure ripple patterns from two musical notes, as in the illustration below, the overall noise has an unpleasant, wobbly "WaWaWaWaWaWa" sound as the volume goes up and down. This effect is known as *beating* or *beats.*

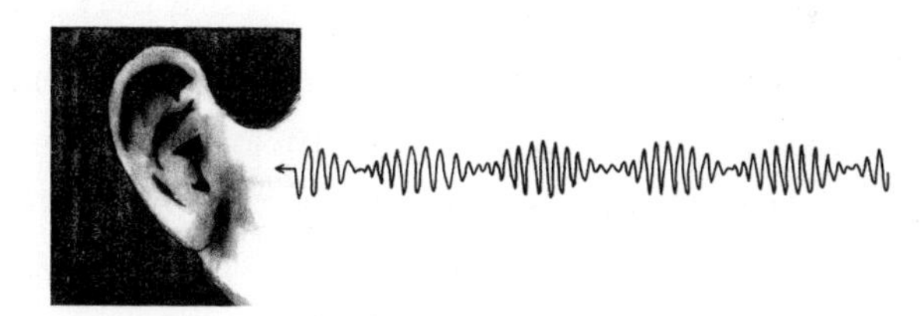

The combination of the ripple patterns of two notes which do not have a simple relationship. In this case, one note has 17/18ths the frequency of the other (which is about the same as the interval between two adjacent notes on a piano). This combination sounds as complicated as it looks. The overall sound is one of two clashing notes and a "WaWaWaWa" effect as the overall volume fluctuates up and down.

The fact that a combination of notes with a complicated relationship sounds rough is the reason why instruments which are out of tune sound awful. In fact, "out of tune" simply means that the nice, simple relationships we prefer to have between notes have been replaced by complicated ones. It doesn't take much de-tuning for the relationships to be ruined in this way; the harmonious sound of two notes an octave apart can be ruined if one of the frequencies is wrong by just a few percent; 110Hz plus 220Hz sounds good, 110Hz plus 225Hz sounds unpleasant by comparison.

One final point to make here is that a musical chord is a combination of three, not just two, notes. But I have only used two notes in my discussion in order to keep things simple. Three notes fit together in a more complicated way than two—but the principles of note combination are the same.

How do we use chords and harmonies?

Rhythm guitarists in rock or pop bands play chords most of the time to provide the harmonies which accompany the melody of the song. Their job usually involves strumming several strings at once to produce a chord, which they repeat a few times before moving on to another one. The notes which make up the chords are chosen to support the notes within the melodies, and this means that the chords and melodies often use some of the same notes. For example, if a certain bit of the tune uses the notes A–B–C–D–E, then a typical accompaniment would be the chord made up of the notes A, C, E. We don't slavishly follow every note used in the tune; we just pick suitable ones which fit. This chord would obviously give most support to the notes within it (A, C, E), so we would use it if those were the notes we were emphasizing in the song. If we had wanted to give prominence to the notes B and D in the same bit of tune, we could have used the chord which uses the notes B, D, F.

You may have noticed that I am not using consecutive letters for my chords. The simplest chords don't involve notes which are right next to each other in the scale because, as I've discussed, notes that are too close together produce harsh combinations. Consecutive notes of a scale are either a semitone or a tone apart in pitch, and I mentioned earlier that notes a semitone apart compete for our attention rather than support each other. The same is true, to a lesser extent, for notes a tone apart, so any consecutive notes from a scale will clash if they are played at the same time. For this reason, a chord made up of the notes A, B, and C, for example, would sound very anguished indeed, as the B would clash with both the A and the C. This sort of chord would not be of much use in accompanying a melody, but it would be right at home in something very tense like "The Devil's Staircase."

The notes in simple, harmonious chords need some breathing space between them in order to support each other, and three alternate notes from whatever scale is being used gives us the commonest type of pleasant combination. However, even in pop songs it is customary to add a little bit of spice to occasional chords by first building a "nice" team of three notes and then adding a single clashing note. So we might use C, E, and G, with a B thrown in to add a bit of tension because it will clash with the C. Our rhythm guitarist (who should really be called the harmony guitarist) provides these groups of notes as a background to the melodies produced by the lead guitarist or singer.

In other musical situations we don't have one person providing the melody and another giving us the harmony. Solo pianists, for example, do both jobs at once, generally playing the melody with their right hand and the chords/harmony with their left. On the other hand, classical music often involves a large team of orchestral players. When an orchestra plays, only a few of the members will be playing melodies at any one time, and the other musicians will play harmonies to accompany them. The composer will often pass melodies around from one group of musicians to another to keep the listener interested. In *Boléro* by the French composer Ravel, the music gradually gets louder as the tune is passed around the orchestra and more instruments join in. The harmonies are kept pleasant and warm until just near the end, where the composer injects a lot of tension for a dramatic final climax.

Chords and harmonies form the background to the melody and also support the punctuation of the phrasing of the music. In fact, you could remove the words and tune from any song and still be able to tell where the ends of the verses were from the harmony alone.

If you are accompanying a melody with chords, the simplest thing you can do is to repeatedly play all the notes of the chord together. Alternatively, you can add an extra layer of interest to the music by playing the notes of the chords one at a time as a sort of continuous, overlapping stream of notes. A chord played as a stream of its individual notes is called an *arpeggio* and this is the basis of the popular folk guitar technique of finger-picking. (Good finger-picking guitar players can play the arpeggios of the chords and a melody at the same time.) Arpeggios add a layer of complexity and subtlety to music because you can choose exactly which notes from the chord will coincide with particular notes in the tune and also add a rhythm to the arpeggio pattern.

Arpeggios are common throughout music and can be found in just about any classical piece, particularly anything with "romance" in the title. The famous slow movement of Beethoven's Moonlight Sonata is a stream of arpeggios with a tune on top, but possibly the best example of a piece made *entirely* of arpeggios is the "Prelude in C Major" by J. S. Bach. There is no real tune, just a series of chords played as arpeggios. Confronted with a piece like this, composed by Bach, most other composers would treat it, quite rightly, as a precious jewel to be admired to the point of jealousy. Not so the nineteenth-century French composer, Charles Gounod. Gounod took one look at Bach's Prelude and thought "A piece made entirely of arpeggios? What a waste . . . where's my book of spare tunes?" The result of this rush of blood to the head was "Ave Maria"—accompaniment by Bach, tune by Gounod—and I have to admit he did a damn fine job of it.

Rock and pop bands also love arpeggios. The opening to "Stairway to Heaven" by Led Zeppelin is a series of arpeggios, as is

"Hotel California" by the Eagles. Status Quo, on the other hand, generally eschew arpeggios and consider them to be a namby-pamby waste of valuable recording time—they revel in rapid repeats of full chords for a high energy effect. Beethoven and Status Quo are in total agreement on this "loud, repeated full chords = energy," as you can tell if you listen to the opening seconds of Beethoven's "Hammerklavier" Piano Sonata or his Fifth Symphony (the one which goes Da Da Da Daah! Da Da Da Daah!).

The most complex type of harmony is called *counterpoint*. "Counterpoint" describes the situation in which you accompany one melody with another melody—in this way you can have two, three or even more tunes playing at the same time. For most of us, the only personal involvement we have in this are those children's songs where two or three singers start the same song after a certain delay—like this:

JIM: Frère Jacques, Frère Jacques, dormez vous? Dormez vous? Soggy semolina...

KIM: Frère Jacques, Frère Jacques, dormez vous?...

TIM: Frère Jacques, Frère...

Or

London's burning, London's burning, fetch the engines, fetch the engines...

London's burning, London's burning, fetch...

London's...

This method of playing the same tune after a certain delay is called a *canon*. The delay means that you are both singing different notes at any one time—which is a similar musical effect to both of you singing different tunes. A slightly cleverer version of the canon involves this idea of singing the same song after a slight delay—but starting on a higher or lower note.

Counterpoint often employs these techniques, but can also

involve different tunes played at the same time. The tunes must work together and, usually, some of the tunes accompanying the main melody are kept fairly simple to prevent the whole thing turning to incomprehensible musical mush. You can't just play any old tunes at the same time because the combinations of notes would occasionally sound dreadful.

Composers have to use a lot of skill to write counterpoint—and a piece which relies on the interplay of counterpoint as its main content is often called a *fugue*. A master of this technique, such as Bach, can organize eight or more tunes playing simultaneously. But this is too clever for us mere mortals—our ears probably can't distinguish more than three tunes at once. If you want to hear some excellent examples of counterpoint, I recommend the Concerto for Two Violins (in D minor) by Bach, and if you want to hear a great fugue, it's best to listen to one played on a solo instrument so you can hear the separate tunes (called *voices*) clearly. Bach's "Little Fugue in G minor" played on a piano is a good example. It starts with a melody played without accompaniment, but before it finishes the same melody starts up again, played on lower notes, and it starts up again a bit later, played on even lower notes. There are other tunes mixed in but they are short and simple by comparison. One distinctive feature of most fugues is that they involve tunes which have an easily recognized beginning, so that each time the main tune jumps into the mix you can hear it clearly.

One or more of the basic techniques of harmony (drones, chords, arpeggios and counterpoint) will have been used in nearly all the Western music you have ever heard—from Bach's concertos to the Sex Pistols. Except for the case of unaccompanied melodies, harmony is almost as important as the tune in Western music.

As we shall see, the Western fascination with harmony meant that we had to develop a peculiarly scientific way of dividing the octave up into twelve equal-sized steps, from which we use a team of seven notes at any one time. The theoretical work began over 2,500 years ago and it only took us 2,000 years to get it right. The final result was the *equal temperament* system, which forms the subject of the next chapter. But before we get into equal temperament, it's interesting to take a quick look at musical systems which haven't followed this harmony-dominated route.

Some differences between Western and non-Western music

The big difference between Western and non-Western music is that only Western music uses chords and harmony a lot in instrumental music. For this reason we have developed instruments which can produce several notes at the same time, such as pianos and guitars. As well as these *polyphonic* (many notes at a time) instruments, we have a large number of *monophonic* (one note at a time) instruments such as flutes and clarinets, but Western music tends to use them in groups so that we can accompany the melody with harmonies. This system of harmony is built on notes which have a strong family link. The relationship between the frequencies of the notes is kept as simple as possible in order for the harmonies to work—which means that the choice of notes is limited.

Non-Western traditional music involves melodies which are allowed to roam much more freely in pitch. This means that the accompanying harmony must be much simpler, to avoid ugly clashes. Imagine that you are part of a five-aircraft acrobatics team. You can either choose to have a limited number of carefully organized moves and all work together, or you can give one pilot freedom to do what he wants while the rest of you stay low as a background attraction. In either case the crowd will be impressed.

Western music has developed in the first of these two ways—the accompanying notes can do a lot of zigzagging around the melody because the number of notes involved is limited, and all the musicians are using the same group of notes. Non-Western music has opted for freedom of melody—and if three or four musicians try zigzagging around each other with this level of freedom, one of them is bound to zig when they should be zagging, and the result would be a mess.

For this reason, traditional non-Western music places far less emphasis on chords and harmony. Indian classical music, for example, tends to use one, or perhaps two, melodic instruments together with percussion and/or rather static, drone-like accompaniments. Good examples of this type of music can be found on most recordings from the Indian subcontinent with the word "raga" or "rag" in the title. "Raga" means color or mood in Sanskrit—and is the name given to an improvised piece of music. Probably the most famous musician of this genre is the sitar player Ravi Shankar, and a great example is "Raga Anandi Kalyan," which he plays with his daughter Anoushka, who is also a world-famous sitar player.

Instrumentalists of this calibre get involved in world tours and, through meeting Western musicians, have been involved in lots of "East meets West" musical collaborations. Some of these are great (e.g., Nusrat Fateh Ali Khan singing "Mustt Mustt" on the album of the same name), some are interesting and some are . . . profoundly lamentable. These collaborations usually involve some sort of compromise between the musical styles and scale systems involved—but the musicians don't care, and the results can be excellent.

The main advantage of non-equal temperament systems is that they can allow more room for emotional expression in the melody. The instrumentalists train for years to swoop up and down around the basic seven notes of the scale they are using. Western blues and rock musicians also spend a lot of time "bending" notes on their guitars and mouth organs, but the big difference is that the final destination pitch of the "bend" is clear to the listeners and other band members. The note will eventually turn out to be one which fits in with the harmony of the accompanying chords. For an

Indian instrumental musician there are many more options available as the final pitch—which is why it would be so difficult to prepare an appropriate harmony. However, over the past fifty years or so, much of the popular musical output from India and Japan has followed the European pattern and uses the equal temperament (ET) system—with lots of chords and harmony.

8. Weighing Up Scales

During your school history lessons you may have learned about armies of rugged men using scaling ladders to get into castles. Similarly, in your French lessons, you probably learned that the French word for "stairs" is "escalier." Both *scal*ing ladders and e*scal*-iers have steps in them which help you rise from a lower position to a higher one—and both words derive from the ancient Latin word *scala*, meaning steps or ladder. *Scala* is also the basis of the word *scale* in a musical context: a scale is a sequence of notes arranged as a series of upward (or downward) steps which take us from one note to another. Generally a scale covers an octave. So we start at a note of one frequency and go up to a note of double that frequency.

There are lots of ways of climbing from one note to the note an octave above—and so there are lots of different types of scales. The most common one we use in Western music is the major scale and I will be explaining what that is in the next chapter. One thing I want to make clear at the moment, though, is the link between the terms "scale" and "*key*." Let's take C major as an example. The scale of C major involves a specific group of seven different notes, but they are only called a scale if you play them one after another as a rising or falling sequence. If you use the same group of notes to produce a piece of music, the melody will be jumping from note to note in all sorts of different patterns and sequences and, as we are no longer merely playing a scale, we say that the music is in the key of C major.

The system of intervals, or musical steps, which produces a musical scale is not set in stone and is different in different parts of the world. Traditional Indian or Japanese music does not use exactly the same intervals as European music—which is why it sounds exotic to Western ears. However, just about all scale systems follow two basic rules:

1. Scales are based on a series of intervals, which are divisions of a naturally occurring interval called the octave.

We have already seen how two notes an octave apart sound good together. The close relationship between such notes means that, in some cases, you can accidentally produce a note an octave above the one you want. For example, if you blow a little too hard into a penny whistle or a recorder, the note you normally get is replaced by another, higher note which is exactly an octave above the usual one (you can also do this with the note you get when you blow across the neck of a bottle). If you twang a guitar string and then touch it gently halfway along its length (above the twelfth fret), the note will change into the one an octave above. Jumps in pitch of an octave occur naturally in birdsong and even in the sound of some squeaking doors, because the octave is linked to the physics of how notes are produced, as we shall see later.

This sweet-sounding, easy-to-produce interval is the basis of all musical scale systems. The octave is, however, a very big musical interval.* The range of an untrained singing voice is usually only a couple of octaves or so. It is therefore no surprise that all musical systems have found the need to divide the octave up into smaller intervals in order that we can have a variety of notes to sing.

2. Musicians don't generally use more than about seven different notes at a time—even if the octave has been divided up into more steps than this.

---

* The first two notes of "Somewhere over the Rainbow" are an octave apart.

We shall see that, although European (or Western) music divides the octave up into twelve equal intervals, it generally only uses chosen teams of about seven different notes at a time. These groups of seven notes are the major and minor keys, the ones used in nearly every piece of Western music you have ever heard. Each team of seven has its own name: for example, the notes C, D, E, F, G, A and B make up the key of C major; and the notes F, G, A, B flat, C, D and E make up the key of F major.

The fact that we only choose to use seven different notes at a time fits in well with research carried out in the 1950s by the American psychologist George A. Miller, who studied the capacity of our short-term memories. After testing people on their ability to remember sequences of numbers, letters and tones, he came to the conclusion that the limit of our short-term memory is about seven items. This limit of approximately seven is also common in other musical cultures. Indian musicians, for example, divide the octave up into twenty-two steps, but they also choose a group of seven notes to be the basis of any particular piece. (Indian musicians also have access to groups of secondary notes associated with their basic group of seven—as we shall see, Western music doesn't use secondary notes because they would interfere with the harmonies we use.)

In Western music the composer or songwriter is not forced to stick to his original chosen group of seven notes throughout a piece of music; it is common to move from one key to another as the piece progresses, to add interest to the music. As there are only twelve different notes to choose from overall, and each key contains seven members, it is obvious that, if we move from one team to another, some of the notes will be in both keys. In fact, the most common key changes (or *modulations*) involve moving to a key which has only one note different from the key you are in.

Many of you will know that the word "octave" comes from the Latin word for "eight" and you will have heard that there are eight notes in an octave. So you might be wondering why I keep referring to seven different notes. Well, if you play or sing a full octave scale you do use eight notes, but the top and bottom ones are an octave apart and, as I said in the last chapter, notes an octave apart are musically very similar and even have the same name. So in an octave we have eight notes, but only seven *different* ones. For example, an octave scale of C major is C, D, E, F, G, A, B, C.

The three main things to remember about scales are:

- we need scales to enable us to memorize tunes;
- scales divide an octave up into smaller intervals—and give us a selection of notes;
- too few notes in play at any one time would be boring, and too many would be confusing.

The way we name the notes and the method by which we choose the members of each team of seven will be discussed in the next chapter. First, I want to explain how we managed to divide up the octave into the twelve different notes from which we pick those teams.

As you can imagine, the ancient societies didn't get out of bed one morning and arbitrarily decide to divide the octave up into twelve steps. Music was developed by musicians who didn't know about frequencies and such things; they just knew what sounded good. In the early days of music, more than 2,000 years ago, things were simpler than they are today. They didn't use twelve different notes in an octave, or even seven. They used five.

The mother of all scales—the pentatonic scale

Although most musical systems around the world now use about seven notes at a time, nearly all of the musical scale systems that humans have ever used—from well before the ancient Greeks to the present day—have been based on a scale which uses only five different notes in the octave: the *pentatonic* scale ("penta" means "five" and "tonic" means "note" in Greek).

A lot of Japanese, Chinese and Celtic music still uses this pentatonic system—and it is also loved by blues and rock guitarists. Typical examples of the five-note system are the songs "Amazing Grace" and "Auld Lang Syne." (If, like most folks, you need a reminder of the lyrics of "Auld Lang Syne," they are "Should auld acquaintance be forgot... la lah la la la lah... lah, happy new year!... hic!... give us a kiss... la lah la la lah.") You will also find a very clear example of a simple pentatonic rising scale in the guitar line at the beginning of the 1960s hit single "My Girl" by the Temptations.

The notes in a pentatonic scale have a very simple mathematical relationship to each other and this makes them form an excellent self-sufficient group—whether they are played in a sequence to produce a melody, or in combinations to provide a harmony.

Next time you are near a piano, try using one finger to pick out a melody using only the black keys—this is the sound of a pentatonic scale or key. All the melodies you play using these five notes to the octave will sound pleasant. Now try playing any two black notes at once—most of the combinations sound harmonious. Only when you play two black notes which are next to each other does the harmony sound clashing and unsettled.

In the most commonly used keys in Western music—the twelve major keys—we have seven different notes in an octave. Some of the notes are only a semitone apart and, as we know, this is the most inharmonious combination for notes played together. Also, as we shall see later, the arrangement of the seven notes in a major key is responsible for the strong sense of punctuation we feel at the end of musical phrases. The reason why the standard pentatonic scale is so incessantly pleasant for harmonies is that it contains no semitones: the "crowded togetherness" of some of the notes is avoided by reducing the number of different notes in the octave from seven to five. Also, if you have only five notes in an octave the punctuation of musical phrases is a little vague. So, with the correct five notes to an octave you have a very supportive, mutually collaborative team with relaxed punctuation—and it's very difficult to make a jarring or unpleasant noise.

To get a pentatonic scale on our piano we have chosen a particular group of five notes from an octave which has been divided up into twelve equal steps. But ancient civilizations didn't know about our modern system of dividing the octave up into twelve semitones and then choosing which group of notes to use, so how did they manage to choose five correctly spaced-out notes for their harps and flutes?

To get a decent sounding scale we want to have a team of notes which have frequencies which are related to each other. The simplest stringed instrument is a harp, and harps have been used by a large number of ancient civilizations. Let's consider a harp with a range of just one octave. The top string needs to have a fundamental frequency which is twice that of the lowest string—because it's an octave higher—and everything is based around this naturally occurring, pleasant interval. For optimum teamwork, the strings between these two should also have frequencies which are related to the frequency of the lowest string: for example, one and a half times the lowest string frequency, or one and a quarter—anything which involves simple fractions.

It's important to realize that ancient civilizations all over the world *independently* developed a system of tuning their instruments to the pentatonic scale. This tuning system had to be based on a naturally occurring phenomenon; otherwise it would not have been discovered by lots of civilizations.

The physical phenomenon in question involves the vibration of strings. Obviously, you only have to twang a string to make it give off its usual note—but it is also quite easy to persuade a string to produce a couple of other notes which are closely related to the first one. This allows us to set up a chain of events which sounds like the plot of a detective story. We get the first note to tell us who their best friend is, and then we get that note to tell us who *their* best friend is—and so on until eventually we end up with a group of notes which all get on well with each other: the pentatonic scale. Once you have managed to tune a harp to a pentatonic scale, you can copy that pattern of intervals for instruments which don't have strings, such as flutes.

This ability to tune a harp to the pentatonic scale is the musical equivalent of the invention of the wheel. It is the cornerstone of musical development, so let's see how it's done...

Let's imagine I am a trainee harpist in ancient Egypt, trying to tune a six-string harp to a pentatonic scale. Someone has described to me how to do it, but this is my first attempt. At this point I have only two skills: I can change the note produced by any string by putting more or less tension on it, and I can tell when two notes sound the same.

"Hang on a minute," I can hear you say, "why are we using six strings for a five-note scale?" Good point. The answer's quite simple, as I mentioned earlier: an octave scale begins and ends with the same note. So, for a full octave scale with five *different* notes we need six strings.

So here we are in ancient Egypt, with no access to tuning forks; this means that the tuning system I use must be self-contained. The only equipment I can use to tune the harp is the harp itself. It sounds a bit daunting, but actually it's quite straightforward.

The strings of my harp are of different lengths—the longer ones are for the lower notes, and I will number them from 1 to 6 with number 1 as the longest string. I start with all the strings slack, and begin by tightening string 1 (the longest one) until it is just tight enough to produce a nice, clear note. I don't need to worry about the frequency of this note because that meeting in London won't decide on a standard pitch for notes for about another 4,000 years. Any nice clear note will do.

Now I need to use this string to help me tune the other strings correctly—but how do I get this string to produce different notes from the one it usually makes?

I use one hand to pluck the longer string while *gently* touching it with a single fingertip from my other hand—and I gradually move my fingertip down the length of the string as I continue plucking. Generally, my fingertip stops the string from making its clear note and all I get is a "thunk" noise—but when my fingertip is in certain positions I get a clear note.

The loudest, clearest note is produced when my fingertip is exactly halfway down the string—and the note produced is an octave above the normal note given off by this string. So all I have to do is match this "octave above" note with the normal note given off by string number 6 and later, when I twang the two strings normally, they will be an octave apart.

Fine, we have the top and bottom notes of our octave organized, but how do we get nice, related notes for the four strings in between these two extremes?

The answer lies in the fact that, if I go back to my plucking/touching routine, I will find that there are other places on the string where I get a clear note rather than a "thunk." As I said, I get the clearest note when my fingertip is resting on the string at the halfway position, but I also get clear, ringing notes if my finger is exactly one third or one quarter of the length of the string from the end.

It's a shameful indictment of the shabby way we live nowadays, but we have to face the fact that most of you won't have an ancient Egyptian six-string harp lying around. On the other hand, you might have access to a guitar or any other stringed instrument—and you can try to find these clear notes for yourself. Guitars are easiest for this experiment because the positions which are one quarter, one third and one half the length from one end of the string are exactly above certain frets (in fact, that's why the frets are in those positions). If you twang the string with a finger resting gently on it above the twelfth fret, then you have divided the string in half—and you get the "octave above" note. If you do the same thing above the fifth fret, you have divided the string into quarters and you will get the note two octaves above the original string. If you pluck the string while touching it above the seventh fret, as I am doing in the illustration below, you have divided it into thirds—and you get a completely new note. This is the only *new* note we have produced because at the "half" and "one quarter" positions we were merely getting notes which are one or two octaves above our first note—which can be considered to be just higher versions of our original note.

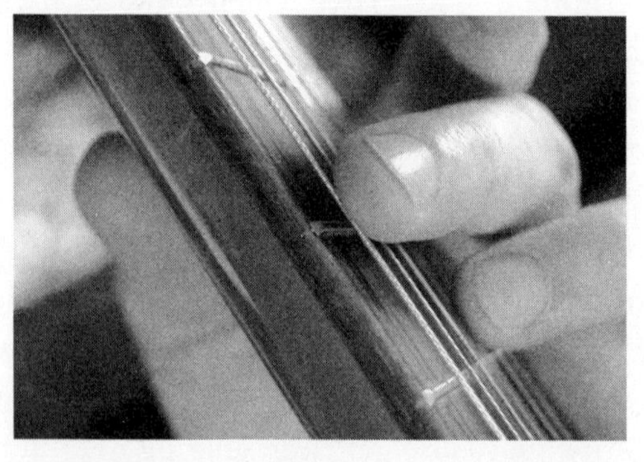

If I pluck a string with my right thumb while touching it gently with a finger of my other hand, I will generally get a "thunk" rather than a clear note. But when my finger is in some positions on the string (as above) I will get a clear note.

The phrase "Twanging the string with one hand while touching it gently with one finger of the other hand" can only be described as cumbersome and, as the note produced sounds like a "ping" rather than a "twang," we will call this method of producing a note "pinging" from now on.*

So—"pinging" can very easily produce three notes from a string:

1. one an octave above the usual string note
2. one two octaves above
3. a new note

You can get other notes by pinging with a bit more skill and effort, but we don't need to worry about them—we will only be using these three.

This ability to produce octaves and one new note for every string is all we need to tune our six-string harp.

So far we have tuned string 6 to match the "octave above" pinged note of string 1.

Then we use the pinged "new" note from string 1 to give us the note for string 4.

Once string 4 has been tuned, we can use its own pinged "new" note to tune string number 2.

String 2's pinged "new" note gives us the note for string number 5.

And, finally, string 5's pinged "new" note gives us the correct note for string 3.

To keep the above description short and sweet I have ignored the fact that the pinged "new" notes will be either one or two octaves higher than the note we want for the next string. But this adds only a minor difficulty to the tuning process. If you would like to know exactly how to do it, I have written out the full instructions in part C of the Fiddly Details section at the end of the book.

---

* Guitarists call "pinging" "playing harmonics."

Excellent: we have managed to tune our harp using only the harp itself—which is the only way it could have worked in the old days. But why stop at only five different notes; why not press on and add another?

Well, there is a reason. If you repeat the pinging process and use the new note provided by string 3, you get a note which is only a semitone below string 6—and this produces a team member which clashes with strings 6 and 1. Nowadays we enjoy an occasional bit of harshness and clashing in our music, but the ancient societies weren't keen on it, so they stopped at five different notes to the scale. (By the way, this pentatonic scale, with no semitones in it, goes by the catchy little name of the "anhemitonic pentatonic scale"—and your task for this evening is casually to include this phrase in a conversation.)

So what is so good about the original pentatonic scale? What does all this pinging, twanging and octaving give us?

Well, when you've done all the math it works out like this:

- string number one has whatever frequency you first chose
- string number two has the frequency of string one multiplied by 1⅛
- string number three has the frequency of string one multiplied by 1¼
- string number four has the frequency of string one multiplied by 1½
- string number five has the frequency of string one multiplied by 1⅔
- string number six has the frequency of string one multiplied by 2.

(Strings three and five are not exactly 1¼ and 1⅔ but they are very close—only about 1 percent off in each case.)

Simple relationships make notes sound good together and, as

you can see from this list, we have a very strong group of simple relationships between these notes, which is why everyone agrees that they sound pleasant. This pentatonic group has been discovered and adopted at some point by just about every human society.

Although the tuning method was probably developed along the lines I have suggested above, it would not have been necessary for every musician to go through all this palaver each time they tuned their instrument. Musicians develop pretty accurate judgment of the size of intervals and would be able to remember this scale. At a fairly early stage of training a harpist would be able simply to tighten the lowest string until it gave a nice clear note and then tighten the others until they sounded right—possibly by remembering a song or two with the correct intervals in them.

Some societies stayed with the purity of this original pentatonic scale, some used it along with alternatives, and in Europe we eventually ended up with a system in which there are twelve different notes to choose from, but we only use about seven at a time. In the next chapter we will concentrate on choosing our various teams of seven. For the rest of this chapter we will be looking at why it took us several centuries to decide how to divide the octave into twelve equal steps. The system we eventually developed to do this is called *equal temperament*.

The European/Western scale system: equal temperament

Why do we use equal temperament?

The Western scale system is the one we use for nearly all jazz, pop, rock and classical music. Its present-day version is called "equal temperament," or ET, and was developed in Europe in the latter half of the eighteenth century. By about 1850 all the European professional musicians were using the ET system and we are still using it today. In fact, it's such a useful system that we will probably keep on using it until we are wiped out by invading Martians (and even the Martians will probably carry on using it for their "light jazz and popular classics" coffee mornings).

Yet, for all its popularity, the ET system is a compromise. And we need a compromise for the usual reason—because we can't have exactly what we want.

What we want is a method of dividing up an octave into twelve steps which also allows the frequencies of all the notes to be related to all the other notes by simple fractions such as $1\frac{1}{4}$ and $1\frac{1}{2}$. This would be great—because all those simple relationships would give us lots of good harmonies. This is the idea behind the "just" scale system—but the "just" system doesn't work for many instruments, as we shall see.

The "just" scale system

The "just" scale system dates back more than 2,000 years and is built upon the idea that simple frequency relationships such as $1\frac{1}{2}$ and $1\frac{2}{3}$ give the best harmonies and that the octave should be divided up in this way. Unfortunately the arithmetic says that this cannot work if you are going to keep the frequencies of your notes fixed at set values, such as the 110Hz we have been discussing. To use the "just" system the musicians must be able to shift the note frequencies up and down a bit during the piece if they want to keep all the chord relationships simple.

Let's take an example of four people singing in harmony. Fred is singing the bass line, and in this song all he has to do is sing the notes A, B, C, A, B, C over and over, while his friends sing lots of other notes. If you listen to the piece, you would think that Fred is repeating himself exactly—but if he did this some of the harmonies would sound horribly out of tune. Using the "just" system,

Fred will have to change the pitches of some of his notes occasionally to make the harmonies work perfectly. His usual "B" could work fine most of the time, but he might need to raise its pitch a tiny bit for certain combinations of notes (let's call this slightly higher note "B★")—so he will be singing

A, B, C, A, B, C, A, B★, C, A, B, C.

The reason for this is that even within the apparently "perfect" relationships of the "just" pentatonic scale there are cock-ups if you look into the relationships between all the strings. For example, string number 4 vibrates at 1½ times the frequency of string number 1, so if the system is fair, strings 5 and 2 should share the same jolly relationship. Unfortunately they don't. If you do the calculation, you find that string 5 vibrates slightly less than 1½ times the frequency of string 2. This means that strings 1 and 4 sound great together, but strings 2 and 5 sound out of tune—so you would have to change the frequency of one of the notes to get your perfect "1½ times" relationship in this case.

Things get even worse if you extend the simple fraction idea to twelve notes in an octave. If you do this, you find that the frequencies of some of the notes need to be changed slightly if you move from one set of seven notes to another, as we do when we change key.

The "just" system can only be used to good effect by instruments which don't have fixed notes—like violins, violas, cellos, trombones and, most important, the human voice. On these instruments it is possible for good players to adjust their notes during the course of a piece so that the combinations of notes always have lovely, simple relationships. However, you can only use this system if *all* the instruments involved are capable of note adjustment. The minute you introduce an instrument with fixed note frequencies, such as a piano, guitar, flute, or almost any other instrument, the "just" system has to be abandoned. This is because, if half the musicians adjust and the others can't, the two groups will be out of tune with each other.

The final outcome of all this is that choirs and string quartets (two violins, viola and cello) tend to use the "just" system to get the best harmonies whenever they are not accompanied by "fixed note" instruments. The "fixed note" instruments need a different scale system, one which compromises on the purity of the harmonies but allows the notes to stay fixed. That compromise is the equal temperament system—and it took us centuries to develop.

The development of the equal temperament system

IN TUNE OR OUT OF TUNE?

We know that people were playing harps as long ago as 2500 B.C. because we have pictures of them doing so on decorated vases from that period. It is more than likely that the practice of music goes back thousands of years before this.

When people moved on from singing to the production of musical instruments like the flute (made of wood or bamboo), they would have come across the problem of where to put the finger holes. The position of the holes determines the pitch of the notes the instrument produces, and it is very easy to produce a flute which has several notes which are not in tune with each other. I'm not talking about "out of tune to modern ears" here, I'm talking about "someone throw a rock at that idiot and chuck his flute on the fire" out of tune—the sort of "out of tune" everyone would agree on.

This sort of "out of tune" describes the situation where some of the notes on a single instrument don't agree with each other. If you took a penny whistle and drilled another finger hole at some random position you would soon hear that your new note was not in tune with the others.

On holiday in Spain recently, I came across one of those street markets which sells earrings to young women, painted plates to middle-aged couples and chickens made of bent coat-hanger wire to all those people who think that life is a hollow sham unless you own a chicken made of wire. Narrowly escaping a vendor of basketwork teaspoons, I found myself in front of a stall selling bamboo penny whistles. When I noticed that someone had simply drilled a row of equally spaced holes along them, I knew that every whistle would be out of tune with itself—you cannot produce a musical instrument that way. Raising one of them to my lips, I played the most unrelated, mournful collection of notes it has ever been my displeasure to hear. Even the local dogs eyed me with thinly disguised contempt. So, of course, I had to buy it—and I present it here for your delectation.

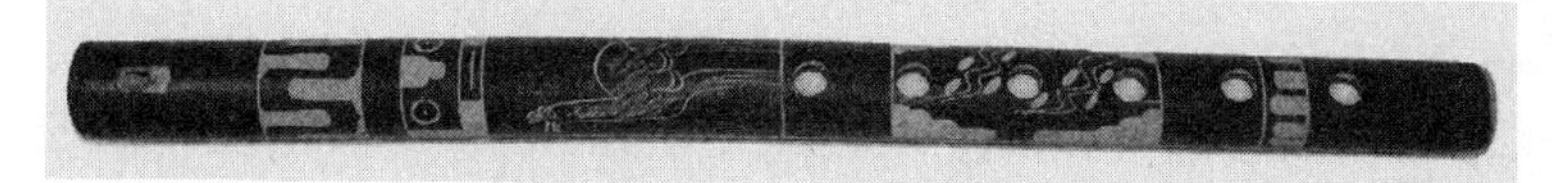

A penny whistle which is out of tune with itself. Someone with no idea about how a scale works has simply drilled a row of equal sized, equally spaced holes along the body of this instrument in the hope that this would create a musical scale. It produces a sequence of notes which have no relationship to each other—so if you try to play a tune on it the jumps between the notes are the wrong size and the effect is somewhere between funny and painful. (If any tax inspectors are reading this, I would like to make it clear that I intend to claim the three euros I paid for this instrument as a tax-deductible expense, to offset against my royalties.)

Anyone can tell when an instrument is out of tune with itself: each of the individual notes might sound pleasant but, played in sequence, they don't form a coherent family. If your harp or guitar is out of tune with itself then you just have to stretch or relax some of the strings until they are back in tune. Strings are always going out of tune because the air temperature or humidity has changed, which is why guitars and violins have tuning pegs (called machine heads on guitars) at the end of the neck. The situation is different if you are making a wooden flute. You can change the notes a little by making the holes larger, but if you put the holes in the wrong place to start with you will probably never manage to get it in tune with itself, let alone other instruments.

The discovery that certain collections of notes sounded good and others didn't inspired the great thinkers of the ancient world to look for some sort of rule from which you could, for example, always put your flute holes in the correct position. Thus began the search for the perfect scale system.

PYTHAGORAS GETS IT WRONG

Although there must have been lots of brainy types who tried before him, our earliest records of an attempt at a scale system which divides the octave up into twelve steps go back to Pythagoras, who lived in Greece in the sixth century B.C. Pythagoras, as you no doubt know, was a mathematician with an unhealthy interest in triangles, who spent a lot of his life trying to develop the perfect shape for Toblerone. Although his triangle research was wildly successful, his attempt at a twelve note scale system was, unfortunately . . . er . . . total rubbish. The idea that Pythagoras developed works well for pentatonic tuning, but it can't be extended to a twelve note system. Before we go over what Pythagoras did, let's make a list of what he (and any other scale designer) was trying to achieve:

1. We want to divide the octave up into smaller intervals (not too many, not too few).
2. We want a system which gives us notes which sound good when they are played together as chords (and, of course, when they are played in a sequence to make a melody).
3. We want a system which allows us to start our tune on any note. If we change our start note, the melody will not change—the whole thing will just be higher or lower in pitch.

Doesn't sound too much to ask, does it? Surely our besandaled hero should be able to come up with a system during his lunch-time stroll down to the beach?

The musicians of Pythagoras' day knew how to tune their instruments—they used thicker or longer strings for the lower notes and adjusted the tension until they had the correct scale, but they didn't know why those notes worked well together. They didn't have a theory for their tuning method, because you don't need a theory to make good music.

We need to bear in mind that Pythagoras was not a musician; he was a scientist—and he wanted to analyze what the musicians were actually doing with their pentatonic tuning and investigate ways of getting even more notes into an octave.

After carefully looking at how the musicians around him tuned their harps to a pentatonic scale, Pythagoras worked out that, if they used strings made of the same material under the same amount of tension, then an "octave higher" string (string number 6) would have to be half as long as the lowest note string (string 1). He then worked out that the frequency of the note produced was directly related to the length of the string involved—if you halved the length of the string you doubled the frequency of the note.

Following this line of thought, he gave the "pinging" method some close attention and discovered that, with identical strings, string number 4 would be two thirds as long as string number 1. "Excellent," he thought, "I'll have this scale thing finished before they call last orders down at the taverna." Going back to his theory about the relationship between frequency and string length, he worked out that a "two thirds" string would give you a note of one and a half times the frequency of the full string note. Once he realized that "pinging" gives you a string which is two thirds as long as the string you started with, he managed to tune his six-string harp using a theory rather than just his ears—bingo!

If only the next bit had gone well he would now be the patron saint of musicians or something, rather than "that bloke with the triangles." Pythagoras had a great deal of faith in simple number ratios and he liked the idea that just by using the ratios ⅔ (for new notes) and ½ (for octaves) he could continue to produce new notes to fill up an octave with smaller steps. In the end, he estimated that he could divide up an octave into twelve steps in this way. He did some rough calculations and it all looked pretty good, but when he took a closer look at the numbers he found that this plan doesn't quite work. If you use this system to divide up your octave, you end up with quite a lot of notes which do not sound good together. Also, you cannot start your song from any note in the scale and get the same result. Both of these problems are caused because—as we climb from the low notes up to the high notes—some of the "steps in the ladder" are different sizes. Pythagoras was so peeved that his cunning plan didn't work out that he took revenge on the world by inventing math exams.

THE "MEANTONE" SYSTEM

People were still hungry for a scale system which gave them more notes than the old pentatonic one, and by now they knew that dividing the octave up into twelve steps was a good idea. Over the next 2,000 years musicians and theorists fiddled about with the unsuccessful Pythagorean system until, by the late 1500s, they found an adjustment to it which just about worked for most of the combinations of notes on a harp or keyboard. This adjusted system was called the "meantone system" and involved using "two thirds and a bit" as its basis rather than the "two thirds" which Pythagoras had set his heart on.

By 1600 the meantone system was in general use for the keyboards of the day (harpsichords, organs, etc.—pianos were not invented until the early 1700s). The system worked well for most combinations of notes but not all—one or two combinations sounded dreadful and were simply avoided by all musicians. For example, in the musical textbooks of the time it was forbidden to play the notes A flat and E flat together because of the horrible racket they made. So the hunt for a theory which would give us a tuning system which worked for all notes continued.

SUCCESS AT LAST

Both the Pythagorean and the "just" systems were attempts to use simple ratios to get an acceptable number of notes in a scale and to have those notes sound good together (because simple ratios make good harmonies). What both of these systems had in common was the fact that, if you kept the math as simple as possible, you would end up with twelve steps to an octave, but unfortunately for these systems the steps were not all of the same size. The two systems also agreed that, once you had assembled your family of twelve notes, some of them were much more closely related (and therefore more important) than others. We continue to use these two principles today: we have twelve notes to the ET octave but we only use seven or so of the most important family members at a time.

Throughout the late 1500s and early 1600s mathematicians tried to solve the problem of producing a scale which worked properly for all the combinations of twelve notes. In the end they cracked it and were, of course, ignored for about a hundred years. Even though Galileo's father (Vincenzo Galilei) got the correct answer in 1581, the equal temperament system didn't come into common use until the late 1700s. The Chinese scholar Chu Tsai-Yu beat Galilei to it by one year, but when he presented his result to the Chinese musical fraternity he got the same response as his Italian counterpart—"get back to the triangles and leave music to the musicians." The British can be particularly proud of the piano firm Broadwood, who refused to change to the new, better system until 1846—hurrah!

Galilei and Chu Tsai-Yu found that calculating the equal

temperament system is pretty easy once you have presented the problem in a clear, logical way. All you need are three focused rules:

1. A note an octave above another must have twice the frequency of the lower one. (This is the same as saying that if you use two identical strings, one must be half the length of the other.)
2. The octave must be divided up into twelve steps.
3. All the twelve steps must be equal. (If you take any two notes one step apart, then the frequency ratio between them must always be the same.)

But you can't just shorten the strings by a certain length each time to get the required result. You couldn't, for example, take a long string of 24 inches and take 1 inch off the length of each of the next twelve strings until you reach the "octave above" string length of 12 inches. You can't do this because it would be unfair to take the same amount off a short string as a long string. That would be like a system where everyone pays $20,000 tax no matter what they are earning.

As the difference between the lengths of the strings must not be a set amount (like 1 inch), what you have to do is make the length of each string a certain percentage of the length of its longer neighbor. Imagine that I am a carpenter. I have an assistant who is paid less than me and a supervisor who is paid more—a perfectly normal situation. Let's say that my supervisor gets $100 a day, I get 10 percent less than that ($90 a day) and my assistant gets 10 percent less than me ($81 a day). Notice that the difference between my pay and my supervisor's is $10 per day but the difference between my assistant and myself is $9 a day—even though we used "10 percent" for both calculations. The reason for the difference is that we took away 10 percent of the supervisor's pay the first time and we took away 10 percent of *my* pay the second time (my pay is smaller than my supervisor's—so the "10 percent" is smaller too).

This is how we use the percentage system to decide on the lengths of the strings; we subtract a certain percentage of the length of any string to get the length of its shorter neighbor. Our mathematical friends Galilei and Chu Tsai-Yu managed to calculate the exact percentage by which each of the strings should be shortened to get a gradual reduction in string length which made the thirteenth string half the length of the first. This calculation is described in part D of the Fiddly Details section, where you will find that the actual percentage involved is that charming, easy-to-remember figure, 5.61256 percent. Thankfully, you will not be required to remember this number. If you wish to show off your extensive musicological knowledge at a party, just remember the 5.6 bit and make up all the other numbers—the trick is to stare them in the eye and sound confident.

The mathematicians produced a scale (the ET scale) with exactly fair, equal steps, which allows us to start our music on any note, and also happens to allow us a lot of good sounding combinations when we play notes together. The interval between any two adjacent notes on this scale is called a semitone and there are twelve semitone steps to the octave.

There is one disadvantage to the ET system but, thankfully, it doesn't interfere with our enjoyment of music. Using the ET system we no longer have the exact simple fraction relationships between the notes which would give us the best harmonies of all. The only interval which has an exact, simple relationship in the ET system is the octave. If you start on a particular note and go up twelve semitones, the note you get will have double the frequency of the note you started with. The next strongest relationship should be that between our original note and the note which should have 1½ times its frequency. In the ET system this has had to be adjusted by a tiny amount. All the other simple relationships have been slightly increased or decreased. But as I said, it doesn't matter much because most of us don't notice these differences, primarily because we have been brought up on the ET system.

However, twelve semitones to the octave gives us rather too many notes for our memories to cope with if we use all of them all the time. For this reason we invented the major and minor scales, which only use about seven of the available twelve notes at one time. This reduction in the number of notes involved makes music easier to remember, and has some other advantages which I will describe in the next chapter.

9. The Self-Confident Major and the Emotional Minor

Mood and music

There are a number of ways in which composers of symphonies, pop songs and car rental jingles can establish or change the mood of a piece of music. Some of these mood effects rely on the animal responses of human beings and some depend on a shared musical culture between the composer and the listener.

For example, when listening to music, we often find an increase in volume exciting. This excitement can prompt the usual physical reactions of increased heart rate and adrenaline production. This is because our subconscious links an increase in sound volume (people shouting, lions roaring) with possible danger.

On the other hand, we associate slow violins accompanying a piano with romance. We do this simply because we have been taught to do so from film music and TV perfume commercials. In turn, we have taught our composers that they need to include some violins and piano stuff if they want us to start getting out the handkerchiefs. In this case there is no real reason for the link other than the assortment of cultural clichés that we are all familiar with—banjos mean hillbillies and accordions mean Paris.

When all is said and done, music is a form of entertainment, so it doesn't matter if our emotional response is "real" (adrenaline) or "learned" (clichés). We enjoy getting the Kleenex out when Pretty

Woman kisses Richard Gere, and we like leaning into the curve when the Millennium Falcon turns to attack the Death Star. Music helps to complete the experience.

Even in the era of silent films there was a mood music industry. Pianists or small orchestras were hired to accompany the action with appropriate music. Sometimes the film came with specially written music, or a list of suggestions of suitable classical pieces. In many cases, however, it was left up to the pianist to improvise while watching the film. You could also buy books full of pieces specifically written to match the moods of any film. These had great names like "Dramatic Tension No. 44" or "Hurry No. 2 (duels, fights)," and my favorites: "Crafty Spy," "Alluring Tambourine" and "Pathetic Love Theme No. 6." I've been involved in a few pathetic love scenes myself—and I could have done with some background music.

Whether it accompanies film or not, the following links between mood and music are fairly reliable:

- We find increases in speed (*tempo*), volume and pitch exciting—and decreases in these three have a calming effect.
- Anticipation is a good mood enhancer, so if the music is quietly repetitive we expect something (frightening or marvelous) to happen soon and the anticipation helps the dramatic effect.
- Music composed in major keys sounds more self-confident and generally happier than music composed in minor keys.

This last point is very important to this chapter, and needs some clarification before we go on to discuss the method by which we build up the collections of notes which are the major and minor scales.

The *major scale* is made up of our old favorite, the pentatonic scale—with a couple of extra strongly related notes added to fill the gaps and take the total number up to seven. This may not seem a big jump in numbers but it makes a surprisingly large difference.

When we perform music we play notes one after the other and/or in groups at the same time—and when we add a couple of extra notes into the mix we get a large increase in the number of combinations available. Think of it this way: if five of you are having lunch together and two of you have to go to the bar to get more drinks, there are ten possible combinations of two people who could go. If there were seven of you, there would be twenty-one combinations of two people. You've only added two people but you've more than doubled the possibilities. It's the same with notes—the combinations available rise rapidly as you add more notes to the group.

The group of seven notes that make up a major scale (or key) are the most closely related group from the original choice of twelve. This makes them sound good and strong together, whether they are played one after the other as a melody, or simultaneously, to give us chords and harmony. As a result of all this solidarity, music played in a major key tends to sound complete and confident. One particular aspect of major keys is that they are very well suited to definite "periods" and "commas" at the end of phrases.

Minor keys involve substituting a couple of the major scale notes for less supportive members of the original gang of twelve—and the resulting music is generally more mysterious and vague, with less definite punctuation. Partly because the music sounds less self-satisfied, and partly because we have been trained to do so, we associate minor key music with sadness and complex emotions. One of the main reasons we link minor keys with sadness is the fact that the lyrics to songs in minor keys portray the whole gamut of human unhappiness, from "My baby done left me" to genuine tragedy, "My printer done run out of ink again."

It's surprising how early we begin to establish our minor key/sadness link. The other day, Herbie, the three-year-old son of a friend of mine, turned to his mother and said, "This is sad music... it's about a cat who got left behind." I have since inspected the sleeve notes of the CD in question and, although Rachmaninoff makes no overt references to disconsolate cats, my young friend does have a point. It will come as no surprise that the music (the first part of Rachmaninoff's Second Symphony) is in a minor key.

This "major key happy/minor key sad" thing is not, of course, an absolute rule. Leonard Cohen, for example, is quite at home being sad and complex in either minor or major keys. And someone obviously forgot to tell Purcell that minor keys are sad before he wrote his triumphant, cheery "Round O" in D minor. It is also worth noting that in Indian traditional music, a scale very similar to our minor scale is associated with happiness and dancing. Generally, however, the existing tradition will ensure that most Western songwriters will continue to write sad songs in minor keys—and the link between sadness and minor keys will continue.

One of the best ways of recognizing the difference between major and minor keys is to listen to a piece which changes from one to the other. The first half of that famous short piece for piano by Beethoven called "Für Elise" is a good example. Like most classical pieces this can be played at a range of speeds because the composers generally give you only a rough indication of speed such as "slowly" (largo) or "walking pace" (andante). In this case we are instructed to play "with a little motion" (poco moto), which I think is almost entirely uninformative—but who am I to argue with a dead genius? If you listen to a version of this piece which lasts about four minutes, you will discover that it opens with a rather sad tune which begins with a pair of alternating notes (dee-dah, dee-dah, dee . . .). This part is in a minor key. This "dee-dah" tune repeats several times and, after about a minute and a half, there are four quick chords and the music changes into a much jollier mood for about fifteen seconds—this is the major key section. The music then returns to the "dee-dah" minor key theme, after which it goes off in a different direction, before finally rounding off with the dee-dah theme.

Another good example of a change from a minor key to a major one is the classical guitar piece "Adelita" by Francisco Tárrega. This piece is less than two minutes long and begins with a sad tune in a minor key, which is then repeated. After this there is a happier interlude in a major key before we finish off with a repeat of the sad, minor key tune.

Moving from one key to another during the course of a piece is called *modulation*, and this technique is commonly used to add interest to the music and alter its mood. Modulation from a major to a minor key (or vice versa) is a good method of mood manipulation, but music also commonly modulates between different major keys, or between different minor keys. I will discuss modulation in a bit more depth once we have established exactly what major and minor keys are.

In the following discussion I'm going to be using drawings of the patented John Powell Ugly Harp to explain my points. This instrument gets its name from the fact that I invented it... and it's ugly. "Why invent an ugly harp?" you might say.... Well, it's not so much a musical instrument—it's more a visual aid. I've assumed that all the strings are made of the same material and that they are all under the same amount of tension. This means that the length of any string is directly related to the pitch of the note it produces. For example, if one string is half as long as another it produces a note with twice the frequency. The harp also has a flat bottom—with the ends of the strings in a straight line so we can compare string lengths easily.

At the end of the previous chapter we had a scale of twelve equal steps from any note to the note an octave above it (the equally tempered, or "ET," scale). This means that we need thirteen strings on our John Powell Ugly Harp to get a complete octave. (We have twelve different notes plus the top note, which is an octave higher version of the lowest note.) Because of the equal size of the steps between the strings we know that you can start your tune on any string and the same sequence of up and down jumps will give you exactly the same tune—the entire tune will just be higher or lower depending on what string you started on. But, as I said earlier, thirteen notes in an octave are too many for our memory to cope with, which is why we came to develop the major and minor scales.

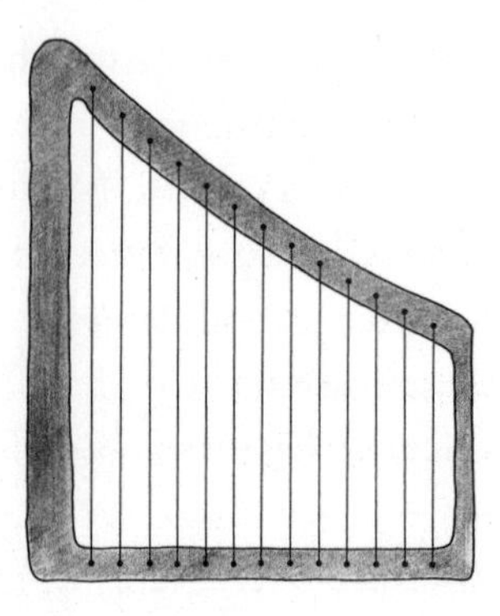

A thirteen-string John Powell Ugly Harp which covers one octave in equal, semitone steps. The steps between the notes are chosen by the ET system.

Major scales

Choosing a family or team of notes for a major scale is similar to choosing a football team from a group of friends (in this case we are going to choose the best seven team members from our group of twelve different notes).

A simple definition of a major scale would be that you take one note* and choose the six notes which are most strongly related to it, to make a self-supporting team of seven. We know that the secret of good harmonies is to use notes with simple relationships between their frequencies so that the pressure ripples join together

* I will call this the *key note*, or team leader, but it is traditionally called the *tonic*, which is based on the Ancient Greek and Latin words for "tone."

to make a regularly repeating, even pattern. We can achieve these simple relationships if the strings on our harp have lengths with simple fractional relationships such as ⅔ or ¾. We have already seen that you can't make a good scale system from these fractions but they do give us the best harmonies.

But our John Powell Ugly Harp is not tuned to simple fractional string lengths; it's tuned using the ET system by taking a certain percentage off the length of each string. "Oh Woe!" you might say, "All is lost!" But do not despair—go and get yourself a calming cup of milky tea and I'll tell you about a handy coincidence.

To create our thirteen-string harp to cover one octave using the ET system, we progressively reduced the length of each string by about 5.6 percent. Luckily this happens to give us a situation where a lot of the thirteen strings are almost exactly simple fractions of the length of the longest string. For example, if we call the longest string number 1, then string number 6 will be 74.9 percent as long as string number 1, which is very close to 75 percent—which is three quarters. The other strings also have lengths which are close approximations of simple fractions. These approximate fractions are so close to the real thing that the harmonies still sound good. For the rest of this chapter I will refer to the string lengths of our harp as fractions of the longest string length. Please remember that I do not mean the *exact* fraction—I am referring to its close approximation, which we arrived at using the ET system.

Fortunately, the ET choice of string lengths on our thirteen-string harp includes six strings which are a very good match for those used to produce a pentatonic scale—and the pentatonic scale is an obvious starting point if we are trying to create a seven-note scale of strongly related notes. So now we can draw our harp with just these notes and see what it looks like. In the illustration below I have labeled each string with its length and frequency as compared to the longest string so you can see that everything is as we want it—only simple fractions are involved.

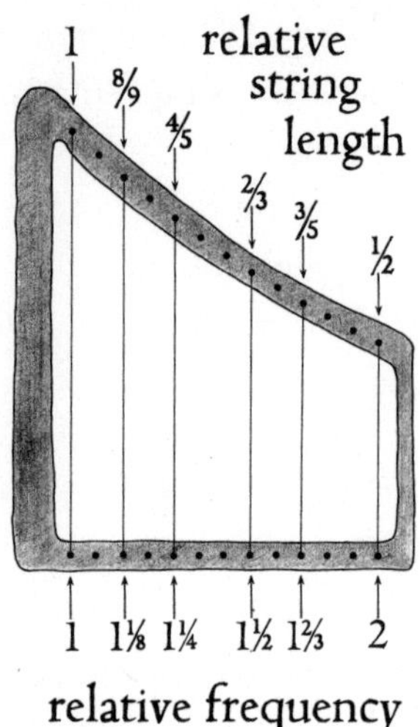

The initial choice of notes for our major scale are the notes of the pentatonic scale. Their lengths and frequencies are shown as a fraction of the longest string.

The pentatonic harp in this illustration looks fairly useful, but both our ears and eyes tell us that there are two big gaps: one between strings 3 and 4 and another between strings 5 and 6. The obvious thing to do to increase the number of notes in our scale is to put one string in each of these gaps—but we have to choose between two possible strings in each case.

Let's look first at the gap between strings 3 and 4. The strongest candidate with the best link with the rest of the group is the longer of the two—because it produces a note which is 1⅓ times the frequency of the longest string.

To fill the gap on the right-hand side we choose the shorter of the two possible strings. It gives a note which is 1⅞ the frequency of the longest string—a good team member—and it also gives us an "almost there" feel to the final part of the rising scale, like this:

String	1	2	3	4	5	6	7	8
	(Home)				(Closest relative)		(Almost there)	(Home again)

Once we add these two strings to our scale, our harp looks like this:

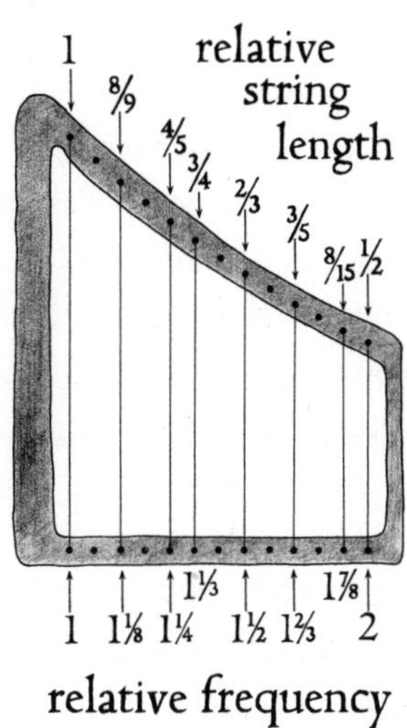

The complete major scale of notes chosen from the original thirteen strings. We have added two more team members to the original pentatonic set.

When the "almost there" note appears in the melody or the harmony it makes a fairly clear demand to get "there," so the listener has a feeling that the next note should be the key note. In fact, this effect is so strong that the technical term for the "almost there" note is the *leading note*, because it leads us on to the key note. Whenever we hear the leading note we build up an expectation of returning home to the key note. This anticipation–resolution effect is used a lot in phrase endings, although sometimes the composer might deliberately frustrate our expectations to make life more interesting. The reason why the punctuation of phrases is vaguer in the case of pentatonic music is because there is no "almost there" note in a pentatonic scale.

I'm using the words "phrase" and "punctuation" here in exactly the same way we use the terms when we are discussing written language. Music has commas, periods, and paragraphs, and uses them in the same way as a storyteller does. The technical term for any phrase ending in music is a *cadence.*

You will notice that there are now only two sizes of gap (or interval) between adjacent notes on our eight-string harp: either the strings are next to each other and therefore a semitone apart; or they are separated by the gap and are therefore two semitones (one tone) apart. Starting from the lowest note in the octave and calling out the names of the intervals between the notes, we would say: Tone, Tone, Semitone, Tone, Tone, Tone, Semitone. Rather than continue writing out this stream of words, I will, from now on, use just the initial letters: TTSTTTS.

To create our major scale we have taken the strongest group of all, the pentatonic scale, and added two more members, one of which (the leading note) helps to strengthen the punctuation of the music. This addition of two members to our team has also given us a tremendous increase in the combinations of notes available for harmonies, without overstepping the barrier of having too many notes for our memories to cope with—a bargain all around, I think you will agree.

The only drawback to the use of major keys is that there is a continuous tendency toward definite, complete statements. Major key music sounds rather self-confident, and sometimes we want the music to be less cocky. In those situations we use minor keys as well.

Minor scales and keys

Although Western music generally restricts itself to major and minor keys nowadays, these two types were selected over time from a larger group of scale systems called *modes*, which used various combinations of tones and semitones to get from one end of an octave to the other. These modes date back centuries but can still be heard in folksongs such as "Scarborough Fair" and they are occasionally used to add a slightly exotic flavor to jazz, classical and pop music. I will discuss modes later in this chapter, but for the moment let's concentrate on modern-day major and minor scales.

The TTSTTTS major scale is actually one of the original modes (it's called the Ionian mode) and it would have been well known as being good for strong, well-organized harmonies because we are using the most closely related team. But strong, well-organized harmonies may be a bit too obvious if we are trying to write dreamy music. In this case we might change some of the team members and use TSTTTST or TSTTSTT, both of which are vaguer. If you write music using these scales the tunes don't come to such an obvious "period" at the end of every "sentence."

By about 1700 most Western composers and musicians had chosen their two favorite types of mode and had just about stopped using all the others which were available. As one of their favorites they naturally chose the strongest team of notes—TTSTTTS—and called it the "major" scale. The other favorite scale, which gave less definite periods, and was therefore suitable for dreamy or sad music, was called the "minor" scale.

But it's not quite as simple as that. For any given major key we always have the same set of seven notes. The really odd thing about minor scales and keys is that we started off with one set of notes, then we altered it by one note. Then we altered it by one note again. "So what?" you might say, "There's nothing odd about that,

John—it's called progress. Things develop and move on—get a grip." If we had moved from the initial set of notes to the next version and then to the next, I would agree with you. But we didn't do that. Somehow it turned out that, for the past couple of hundred years, we have been using all three types at once. And I don't mean we use this one for this piece and that one for that piece. We use all three types of minor scale within a single piece of music. We use the original one if the tune is descending; one of the others if the tune is ascending; and the third one to make up the accompanying chords.

If music was controlled by scientists this sort of untidy nonsense would be forbidden. But music is organized by musicians, with their unkempt hair and faraway expressions. Musicians eventually settle on what sounds best—and they decided that minor keys are more emotionally effective if they change a couple of notes depending on whether the tune is rising or falling. This "deciding" process didn't take place suddenly one night in the pub—it took centuries, as the other options were discarded one by one.

So let's have a look at these different minor scales.

The natural minor scale

Like the major scale, the natural minor scale is one of the ancient modes. It's called the Aeolian mode and you can read a little more about it later in this chapter. It has the pattern TSTTSTT, which means that, compared to the major scale, three of the strings on our harp have been moved down one position, to the next lowest note. As you can imagine, this substitution of three of the strongest team members weakens the team quite a bit. You can see the new arrangement in the illustration below and compare it to the major scale. Two of the original "pentatonic team" have been replaced and the "almost there" note is also no longer with us. The newcomers still make a very pretty noise, but the team lacks the oomph it had, particularly at the ends of phrases.

Still, the whole point of having these alternative sets of notes is to have a different flavor. We don't actually want much oomph from our minor keys. When you're singing songs about harvests failing, or your disappointing loft insulation, you don't want every verse to end with a cheery full stop. Sometimes we need some good, solid ambiguity.

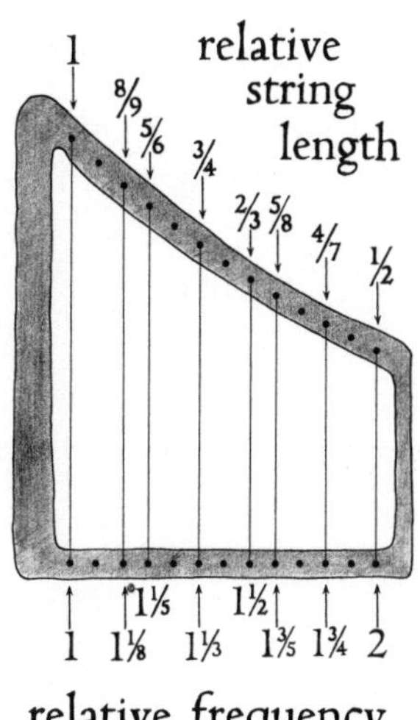

To create the minor scale, three of the notes (string numbers 3, 6 and 7 counting from the left) have been lowered by one semitone. This gives us a less cohesive group.

The natural minor scale can be used alone to improvise or compose pieces of music, but it is usually employed as one of the team of three scales used in most minor scale music.

This natural minor scale was found to be just right for the parts of melodies which were descending in pitch, so it is also called the descending melodic minor scale.

For tunes which were rising in pitch, composers found that they really missed that "almost there" feeling of the next-to-last note of the major scale—and also found that this note was very useful for harmonies. So they developed the ascending, or rising, melodic minor scale.

The ascending melodic minor scale

In the ascending melodic minor scale, two of our natural minor scale notes are reinstated back to their major scale positions. The next-to-last note is returned to its original "almost there" pitch and its neighbor follows it back up to prevent there being a big gap in the scale. You can see what has happened in the illustration below.

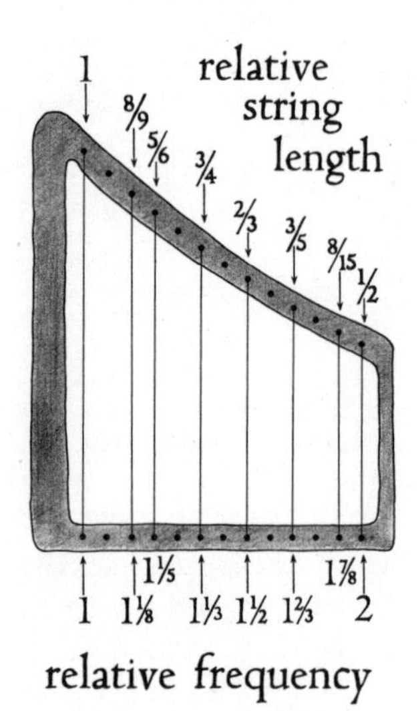

The ascending melodic minor scale. This has only one note different from the major scale (string number 3 has been lowered by one semitone).

So now we have the complete set of notes for melodies in a minor key. We use this scale for ascending parts of melodies and the natural minor scale for the descending parts.

But now, of course, we have a bit of a problem—because we need chords which will suit both the rising and falling parts of our tunes. For this reason a compromise scale was developed from which we can pick our harmony notes. Not surprisingly, it's called the harmonic minor scale.

The harmonic minor scale

The harmonic minor scale is, as I just suggested, a compromise between the descending and ascending minor melodic scales. In this case we take the descending melodic, or natural, minor scale and return only the next-to-last note back to its major key "almost there" position (as you can see below).

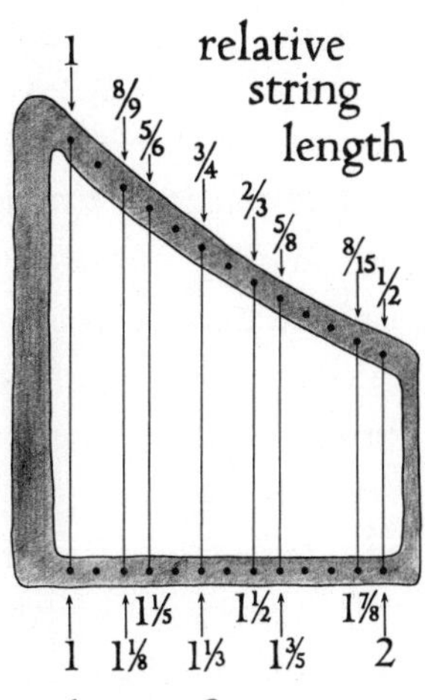

The harmonic minor scale: a compromise between the two melodic minor scales, used for chords and harmonies in minor key music.

So there we have minor keys. A weaker relationship between the various members of the team of notes results in a more complex musical experience. The music sounds less self-confident than music produced in major keys and generally we have come to associate this comparative vagueness with sadness or expressions of deep emotion.

Major and minor chords

The simplest chords consist of three notes taken from the musical scale. As I said earlier, we get harsh combinations if we choose notes which are next to each other, so most chords use alternate notes from the scale.

The most common chords include a basic note (which gives its name to the chord) and its most closely related team member—the one which has a frequency which is $1\frac{1}{2}$ times that of the basic note. If we start with string 1 of a major scale then the $1\frac{1}{2}$ times note is string 5. This is also true of a minor scale, because string 5 does not move when we change to a minor scale. (The fact that the $1\frac{1}{2}$ times frequency note is the fifth one you come to in either scale is the reason why musicians say it is "a fifth" above the key note. The usual technical name for this second most important note in the scale is the *dominant*, because it dominates the tunes and harmony, along with its close relation, the key note, or *tonic*.)

The third note we choose for a simple chord needs to go between strings 1 and 5 but we shouldn't choose strings 2 and 4 because they are right next to one of the others and they would clash with it. So we choose string 3. If we choose strings 1, 3 and 5 from a major scale, we get what is called a major chord and, if we choose the same strings from a minor scale, we get a minor chord.

So a major chord is a frequency we choose, plus $1\frac{1}{4}$ times that frequency plus $1\frac{1}{2}$ times that frequency. In a minor chord we exchange the $1\frac{1}{4}$ for $1\frac{1}{5}$, which is less strongly linked to the other two notes.

As we know, every note involves a series of harmonics and in closely related notes some of the harmonics of one note will be the same frequency as certain harmonics of the other. For example, our old mate A_2 has harmonic frequencies which are multiples of 110Hz, and its strongest team member, E_3, has harmonic frequencies based on 165Hz. Two times 165 is 330, which is the same as three times 110. This means that the third harmonic of A is the same frequency as the second harmonic of E. There are many other matches of harmonic frequencies between these notes, and that's why they sound good together. Using this type of matching we can show that the 1¼ frequency in the major chord is a more supportive match for the other two notes than 1⅕, so the minor chord notes form a less self-confident-sounding team. Once again, as with all things to do with a minor key, we have come to associate this lack of self-confidence with sadness.

A chord is a combination of *any* three or more notes. The major and minor three-note chords we have just discussed are the simplest and most harmonious-sounding ones. Minor chords sound less confident than major chords, but they are stronger and more confident-sounding than a lot of the other chords we can get by putting three or more notes together. For example, by adding "clashing" notes to simple major and minor chords, we can produce chords which sound more colorful, interesting or tense. There are also lots of possible chords which don't include the strong 1½, 1¼ or 1⅕ team members. More complex chords like this help to add movement to the music because they do not sound relaxed or final. Our ears tell us that there must be more steps in the journey before we get to the end of the phrase. When we eventually get to the end of a phrase the music is likely to relax into a simple major or minor chord.

Naming notes and keys

At the end of chapter 1, I mentioned the fact that notes have names which consist of one of the first seven letters of the alphabet and sometimes these letters are followed by the words "sharp" and "flat." Back in chapter 1 I asked you simply to accept this system and not worry about it, but now it's time to look into it properly because we can't discuss our next subject—moving from key to key—without referring to the names of notes and keys.

The main reason we have names for notes is so that we can teach and discuss music. Although there has always been a tradition of passing on music by simply copying what someone is singing, this doesn't work too well for complicated music if you want it to be replicated exactly.

The Western system of writing music and naming notes began with monks who wanted to record their masses and hymns. They needed to make everything easy to remember so they didn't want to use *all* the letters of the alphabet as names for the notes—they chose to use only the letters A to G. Any two notes an octave apart were given the same letter because they are so closely related as far as our ears are concerned—although numbers and other techniques were used to identify which D or E you were talking about (D_1, D_2, D_3, etc.).

So—seven letters to name all the notes in our major scale and the last note would have the same name as the first note but be written slightly differently, or numbered, like this for the scale of C:

C_1, D_1, E_1, F_1, G_1, A_1, B_1, C_2, D_2, E_2, F_2, G_2, A_2, B_2, C_3, etc.

Well, this is all pretty clear so far. Let's draw a two-octave John Powell Ugly Harp with the longest string as C, putting all of the twelve notes we discussed in chapter 8 into each octave.

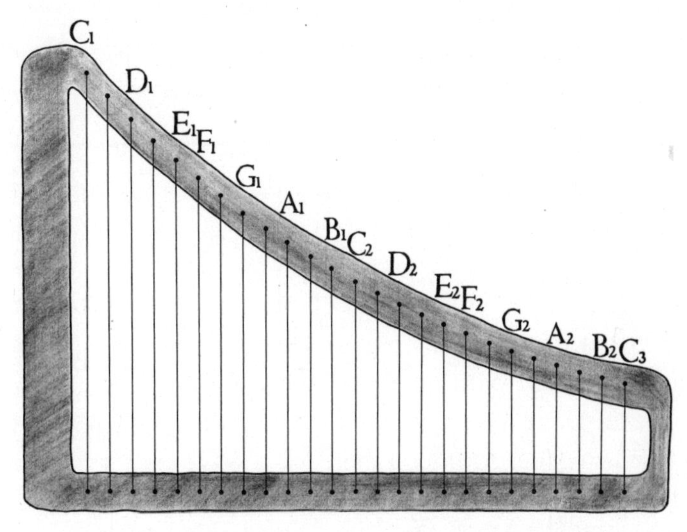

A two-octave harp showing the notes of the C major scale (the longest string is a C).

This illustration introduces us to something which is rather odd and needs some explanation—we have used up all our letters but we haven't named *all* the strings (this is a pretty obvious result if you remember that we only have seven letters but we have twelve different notes to the octave). For example, the string between C and D has no name in the illustration. What do we call such "in between" notes? Well, unfortunately, these notes each have two names. We can refer to them as being higher than the note below them by using the word "sharp," or we can refer to them as being lower than the note above them by using the word "flat." When musicians write these names they usually use the symbol "#" to indicate sharp and "♭" to indicate flat. So "F sharp" is F# and "B flat" is B♭.

In the illustration on the next page I have labeled all the notes—giving the "in between" notes both their flat and sharp names.

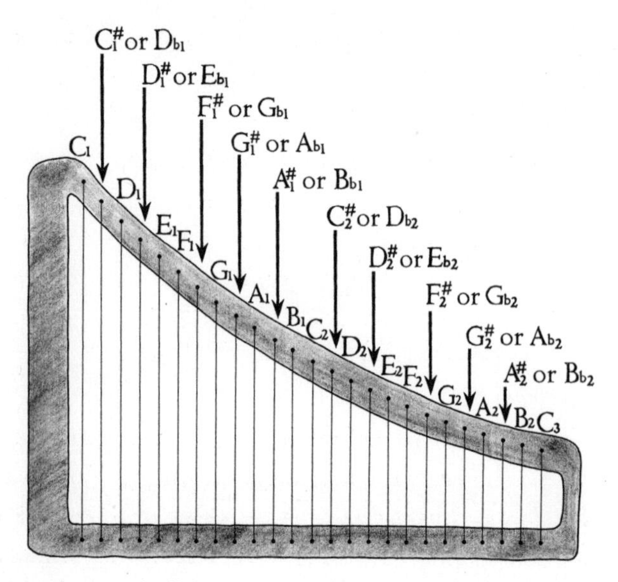

A two-octave harp with all the notes named. Certain notes have both "sharp" and "flat" names.

If you have a spare minute you can pick any note and count up the strings (from long to short) with the TTSTTTS pattern to identify the notes in any major scale. . . . On the other hand, life is too short for this sort of shenanigans—and you didn't buy this book as an activities kit—so I'll do an example for all of us.

If we start the scale with A, we go up one **T**one to B, up another **T**one to C#, up a **S**emitone to D, etc., and the notes we get are the A major scale:

A—B—C#—D—E—F#—G#—A

And it's as simple as that really: pick any of the twelve different notes as your key note or team leader—and the TTSTTTS system will identify who your team members are to give you the strongest, most closely related group—your major key.* (I have written out the notes for all the major keys in part E of the Fiddly Details section at the end of the book.)

You can, of course, do the same for a minor key using the appropriate "T, S" pattern: for example, the melodic minor rising scale is given by the pattern TSTTTTS, and if you start off on E you will get:

E—F#—G—A—B—C#—D#—E.

Of course, the sharp/flat notes are no different from the simple letter notes—they are all equally important—it's just a naming system that has been passed down to us. One of the historic peculiarities about the system is that the only major scale which doesn't include a sharp or flat note is C major. This makes the key of C major look important in some way—but it isn't; it's just the way the naming system evolved.

---

* Even though the sharp/flat notes have two names, it would be considered very peculiar to name the notes of the "A" major scale using "flat" names like this:

A—B—D♭—D—E—G♭—A♭—A

This way of naming notes in a key is not used because some letters appear twice ("A" and "A♭," "D" and "D♭"), which is confusing. We use only either "flat" or "sharp" names for each key—and choose whichever system uses all the letters, like this:

E Major: E, F#, G#, A, B, C#, D#, E

B♭ Major: B♭, C, D, E♭, F, G, A, B♭

Changing from one key to another: modulation

It is clear from the last section that we have twelve major keys and, as I will explain later, they are all emotionally identical: one key will merely have the same pattern of notes moved up or down a bit in pitch. So let's return to the question of why we have so many keys.

Composers and musicians are in a continual battle against boredom—not their own boredom, but the boredom of their listeners. They know very well that if they bore you their income will drop and their children will starve—or, at least, they won't be able to go out for a burger on Wednesday. Music is a form of entertainment and so it has to stimulate the emotions—from jollity to fear (and if you think fear is a bit of an extreme claim for music, you haven't seen the shower scene from Hitchcock's film *Psycho*).

One of the ways a composer can keep up the interest level of his listeners is to change key—from one set of seven notes to another. If this happens, one or more of the notes is changed, and the listener can tell that the team leader of the group has also changed. This team analogy is particularly helpful here: imagine you are the manager of a team whose style of play becomes a bit stale during the first half of a game. At half-time you can put extra life into the team by exchanging a couple of players for substitutes and asking the team to vote in a new captain.

So now you have a slightly different team with a new team leader—which is exactly what happens if you change key in the middle of a piece of music. You might think that a non-musician would not be able to spot the change in team leader (or *key note*) but, in straightforward Western music such as pop, rock, folk, blues and most of the classical music written between 1700 and 1900, the key note is fairly easy to spot, even for an untrained listener. In more complicated music, such as modern classical or jazz, the team leader, and therefore the key of the music, may shift about every few seconds or may be deliberately hidden. In this case, the sense of a key becomes confused or lost.

When we are listening to a piece of straightforward music we identify the key note in two ways, although you probably won't realize or notice that you are doing it. First, a song or any other piece of music is divided up into phrases and the key note will often be the final one of a phrase. If you play just about any pop song—even one you haven't heard before—you will be able to hum the key note within a minute or so. Just pretend that the tune is ending and hum the note it should end on—that will almost certainly be the key note. Our old favorite "Baa Baa Black Sheep" does this: it hits the key note on the word "full" and the final word, "lane."

The other clue which helps to identify the team leader is how often the various notes of the scale occur as the melody progresses—and here we come to an example of musicological fortitude above and beyond the call of duty. Brett Arden of Ohio State University spent many months checking thousands of melodies (more than 65,000 melody notes for the major keys and more than 25,000 for the minor keys) to find out exactly how often each of the notes in a scale occurs. For example, if we number the notes from 1 (the key note) to 7 (the "almost there" note), he found that, in major keys, note 5 occurs most frequently and will be played about four times as often as note 7, the least common member of the group. There are other relationships which hold true for most tunes. For example, in a major key, notes 1, 3 and 5 make up almost 60 percent of the head count of notes in a tune. Your brain recognizes these proportions and this helps us to tell which note the key is based upon. Obviously, you are not aware of your brain analyzing these relationships. You just pick up these clues subconsciously, as you do when you assess which of the guilty-looking eight-year-old ruffians in your garden just kicked the football through the kitchen window...

The most common type of modulation is to change from the key you are in to a key which contains only one different note.

For example, we could be playing away in the key of C major, which contains the following notes:

C, D, E, F, G, A, B.

And we could easily shift over to the key of G major:

G, A, B, C, D, E, F#,

which has the same notes, except the F has been changed for an F sharp.

If we do this, the music receives an emotional lift because one of the notes has been raised. We also get extra (subconscious) interest because the key has changed and we can sense the change of team leader—from C to G.

On the other hand, we could change from C major to F major, which contains all the same notes as C except for the fact that the B is taken down a semitone to B flat. In this case we often get the impression that the emotional intensity of the music has switched down a gear, although we still get the enhanced interest from the change in team leader.

This "up a gear" or "down a gear" effect has nothing to do with the actual properties of G major or F major—the different keys have no intrinsic emotional shading. It's the process of change which gives us the emotional impact, and the effect fades off quite rapidly (within a few tens of seconds). Imagine yourself standing in a big hamster wheel—it has been stationary for a while and you're bored, so you take a single step forward onto the next rung. Everything gets a lot more interesting for a little while, but soon the step you moved to becomes the one at the bottom of the wheel and it's all boring and stationary again. Stepping backward onto the rung behind you has a slightly different effect—but it's still transitory. The rungs are identical: it's the changeovers which are interesting. If you want to keep life stimulating you are going to have to keep changing rungs.

Composers occasionally inject a surge of interest by shifting several rungs at once—to a key which has a lot of different notes in it—from C major to E major, for example. Ravel does this as a dramatic flourish near the end of his *Boléro*. But most often, keys change to a *neighboring key* (one with only one different note).

Modulating a repeated phrase to a key that is a semitone or a tone above the one you start in (shifting up from B major to C major, for example) never fails to brighten the music because it feels like a change in gear, which is why it has become known as the "truck driver's gear change" or "truck driver's modulation." The technique also revels in the name "the cheese modulation" ("cheese" being the general name given to pop music which has passed its "best before" date).

This modulation is commonly used to give a sudden lift in energy to pop songs, particularly in cases where the chorus is repeated a lot. "I Just Called to Say I Love You" by Stevie Wonder uses this technique a couple of times, but the most notable example occurs when the title of the song is repeated, three and a half minutes into the track. Another very effective example of this type of modulation can be found in "Man in the Mirror" by Michael Jackson. In this case, the key change occurs (as Michael sings the appropriate word) two minutes and fifty seconds into the song.

If a modulation involves movement between two major keys or two minor keys, any change in mood will be short-lived, because the effect is linked to the action of changing—you are just changing rungs on the hamster wheel. If, however, you change from a major key to a minor one (or vice versa) the change in mood will remain in place. This is because, although the effect will be strongest just after the change, you have genuinely moved from one musical landscape to another—like jumping from a steel hamster wheel to one made of wood. Changing from a major key to a minor one will result in a more complex, emotional or sad mood and a move in the other direction will make the music sound more determined and self-assured.

If you change key very often (as some jazz and classical composers do), then the listener may become rather confused and the music will sound a little unstable. If, on the other hand, you don't do this often enough (like some pop bands), the music can become very predictable and bland.

The same team of notes can sound minor or major if you change the team leader

If we consider the simplest version of a minor key, the natural minor, we can construct either a minor or a major key out of one set of seven different notes.

For example, the scale of C major is:

C—D—E—F—G—A—B—C

and the scale of A natural minor is:

A—B—C—D—E—F—G—A

This is the same group of notes except that, for the minor key, we have As at the top and bottom of the scale instead of Cs.

We can pull the same trick with pentatonic keys. The pentatonic major scale which starts on C is

C—D—E—G—A—C

This uses exactly the same group of notes as the pentatonic minor scale which starts on A:

A—C—D—E—G—A

Yes—I agree—this sounds bonkers. How can the same bunch of carefully chosen notes be either C major or A minor?

But it's true—using the same set of notes, you can give a minor (sad, reflective, weak full stops) or a major (happy, positive, strong full stops) feel to the music simply by changing the team leader.

"Aha!" you say, "but if I'm hearing exactly the same notes, how do my ears know that the team leader has changed?" Well, as I said earlier, the team leader, the key note, is fairly easy to spot in straightforward music. It's really all about emphasis—and we are all used to small changes in emphasis making a big difference to what we say. For example, the following two sentences have very different meanings because, although I have used all the same words in the same order, I have changed the emphasis by moving the comma, which makes one sentence insulting and the other congratulatory.

"I'm not a fool like you, I spend my money wisely."
"I'm not a fool, like you I spend my money wisely."

So if we take the notes of the key of C major (C, D, E, F, G, A, B) and use C, E, G as our favorites, particularly C at the end of phrases, we will hear the music as having a major key feel. If we use the same notes, but make A, C and E our favorites, particularly A at the end of phrases, then the music will have a minor feel. The "team" analogy comes in useful again here: if you took any soccer team and changed the roles of some of the players (by giving the goalkeeper the center forward's job, etc.), then the team would perform with a different style even though the players were the same.

If you have access to a piano you can check this for yourself. Try making up simple tunes using one finger and playing only the white notes. If you end all your phrases on the note C, the music will sound fairly cheery and strong. If you stay on the white notes and end all your phrases on the note A, then the music will sound vaguer and sadder.

Choice of key

Composers need several keys, so they can, if they wish, hop from one to another during the course of a piece. But what makes them decide to start a piece in a particular major key if there is no mood difference between any of the major keys? And, similarly, why would they choose this minor key instead of that minor key?

Well, there are a number of reasons to choose a particular key to start your piece—but none of them have anything to do with emotional content. These reasons can be divided into five categories: instrument design; range; composer delusion; perfect pitch; and "first come first served."

Instrument design

A lot of pop songs and most classical guitar pieces are written in the keys of C, G, D, A and E because these are the easiest keys to play on a guitar—the chords and tunes in these keys allow you to get the maximum reward for the minimum effort. For example, if you are a total beginner, I could teach you to play the chords to a simple pop song written in G major in about fifteen minutes, and you could accompany a singer with a fairly shabby version after about three hours' practice. If we moved the key one semitone up—to A flat—or one semitone down—to F sharp—it would take you ten times as long to get even a shabby version together. This is because the finger positions for the chords are a lot more difficult as a result of the way guitars are tuned. Lots of other instruments are also easier to play in some keys than others, and the choice of key is driven not by musical considerations, but by ergonomics. For example, the sort of "big band" which accompanied singers like Frank Sinatra in the 1950s and 1960s involved trumpets, clarinets and trombones, which are easiest to play in the key of B flat, so a lot of those songs are in that key.

Range

Every instrument has a top note and a bottom note. You have to make sure that the piece you are writing fits on the instrument—and this could affect your choice of key. For example, the lowest note on a flute is middle C, so there won't be much solo music written for the instrument in the key of B because it's useful to have the key note close to the bottom of the instrument's range.

Also, the key of a song may have to be raised or lowered to suit a particular singer's voice.

Composer delusion

Many composers, like lots of other musicians, suffer from belief in a myth I will discuss later in this chapter, which suggests that certain keys have specific moods. So they write gloomy music in A♭ and jolly music in A because they think that those are the most suitable keys.

Perfect pitch

If you are a composer with perfect pitch you will hear specific notes in your head every time you think up a tune. Whatever key it appears in will probably be the one you'll use when you write it down.

First come first served

If, like most composers, you have not got perfect pitch and you quite often come across useful tunes when you are fooling around on a piano or some other instrument, you will probably stick with the notes in the key you first found them in, unless there is a good reason to change them. Some composers have favorite fooling-around keys. A famous example of this is the songwriter Irving Berlin (who wrote Bing Crosby's hit "White Christmas" and "Let's Face the Music and Dance"). By his own admission, Irving Berlin was a dreadful pianist, so he played and composed nearly everything in the key of F# major, a key which falls under the fingers very nicely (it uses all the black notes on a piano together with only two white notes). He couldn't write music so he paid musicians to watch his fingers and write it all down, and later they would, if necessary, change the key to suit any instruments or singers involved.

Modes

Although our major and minor keys are a development from the ancient pentatonic scale, we have not yet discussed the story of how we got from one system to the other. It all began with the ancient Greeks, who developed a selection of different scales, all involving seven different notes in an octave, which they called *modes*. The history of modes is a subject of Byzantine complexity, which makes me glad I'm not a historian. Basically it goes like this:

Sometime before 300 B.C. the ancient Greeks used different scale systems of tones and semitones to divide the octave up, and named them after the peoples and territories of Greece and its neighbors. For example, the Lydian mode was named after the area called Lydia, which is now called Western Anatolia.

The Christian church developed a method of singing called Gregorian chant from about 750 onward, which used seven different scale patterns. The church gave these modes the names of the Greek modes without worrying about whether they were the same as the Greek ones. So we now have Christian modes with random Greek names.

The Holy Roman Emperor Charlemagne decided to improve the popularity of these Christian modes by threatening the clergy with death if they didn't use them. The modes became *very* popular.

The modes were used successfully for several hundred years but some of them became less fashionable than others. Eventually, by about 1700, most music was using only two of the original seven—and these two became known as the major keys and the minor keys.

Today, modes are used less frequently than the major or minor keys but you will find them in folk music and some jazz, classical and pop music (when the composer wants the music to have a slightly unusual flavor). Just like the major keys, each mode consists of a set pattern of tones and semitones as you go up and down the scale.

So how do we pick our team of notes for modes? Surprisingly enough, all the modes use the same team as a major scale—but they don't use it in the same way. This sounds a bit weird, but let's go back to our old friend, the Ugly Harp. The next illustration is a two-octave harp showing the notes of the C major scale, and I will now demonstrate how all the modes can be played using the same notes as the C major scale.

We already know how to play the scale of C major on this harp—we just start with the longest string and pluck the C scale notes one after the other. But now we want to play a mode using the same notes. There are seven modes and they have the following names: Ionian, Dorian, Phrygian, Lydian, Mixolydian, Aeolian and Locrian. (Yes, they *do* all sound like heroines from *Lord of the Rings*—except for Mixolydian, who is obviously one of those elvish bowmen with the fancy helmets... but I digress...)

All these modes can be played on a harp tuned to a major scale like the one above. The major scale and all the modes use exactly the same team—the only difference between them is the choice of team captain.

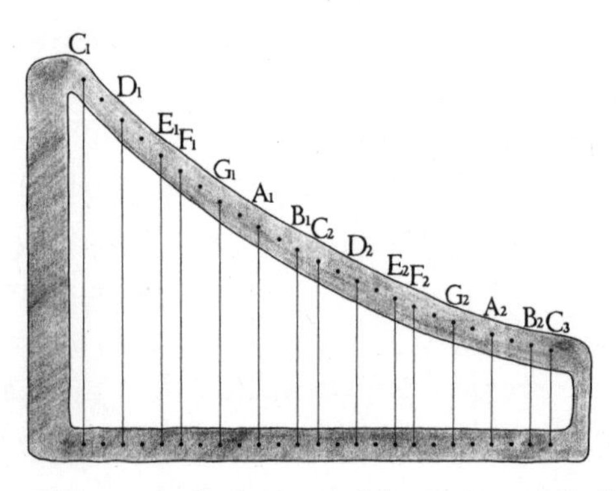

A two-octave harp showing only the notes of the C major scale (the longest string is a C).

We have seven different notes in our scale of C major (C, D, E, F, G, A, B), and seven modes. Each mode uses a different start note (team captain), so one of them must start on C—exactly as our C major scale does. The mode which does this is the Ionian, which is why we never hear of this mode; it's the one we chose to be our modern major scale, so that's what we call it. The other six modes use the following notes as their team captain if they are using the notes from a C major scale:

Dorian mode	D
Phrygian	E
Lydian	F
Mixolydian	G
Aeolian	A
Locrian	B

To play the Dorian mode scale on our harp, we would start and finish on the D strings so instead of the scale of C major, CDEF-GABC, we would play DEFGABCD, and play tunes and harmonies which kept returning to D. To Western ears this sounds a little odd because, given that group of notes, our ears keep expecting tunes to end on C or G. However, once you get used to it, the Dorian mode makes a nice change.

The basic feel of the Dorian mode is similar to a minor key. This is not really surprising because the only difference between the Dorian mode played from D to D and our modern key of D minor (natural) is that the B is lowered to B♭ in D minor—all the other notes are the same. The Dorian mode is used a lot in Celtic music and is also the basis for such songs as "Scarborough Fair" and The Beatles' "Eleanor Rigby."

As I said earlier, the major keys all have a TTSTTTS pattern of intervals as we rise through the scale. If we play a Dorian scale, we play from D to D and this gives us an interval pattern of TSTTTST. As long as you use this interval pattern you can start on any note you like (using an instrument with all the twelve different notes on it) and you will be playing a Dorian mode. You could, for example, start with an A and play the notes A, B, C, D, E, F#, G, A and that would also be a Dorian mode (just as you can start from any note to produce a major scale if you keep to the correct TTSTTTS interval pattern).

If two musicians get together to play a song written in a major key which they both know well, they won't say "let's play it in major" because that doesn't tell them very much. They'll say "let's play it in G major" or "let's play it in E major" and off they go. In the same way, if they wanted to play "Scarborough Fair" they can't just say "let's play it in Dorian"—they must identify which note the Dorian mode they are going to use begins with. They could say "Let's play it in D Dorian" and use the notes D, E, F, G, A, B, C, D but they could just as easily say "let's play it in G Dorian" (G, A, B♭, C, D, E, F, G) or "B Dorian." Just like major keys, there are twelve different Dorian modes.

If you pick the notes from any major scale and make the second note of that scale the team leader, then you are playing in the Dorian mode. To play in a different mode, you choose a different note from your major scale as your team leader:

For the Phrygian mode the third note of the major scale is the team leader.
For the Lydian mode it's the fourth note.
For the Mixolydian mode it's the fifth note.
For the Aeolian mode it's the sixth note.
For the Locrian mode it's the seventh note.

The Lydian and Mixolydian modes are very similar to major keys—in each case one note from the major scale has been moved one semitone. Music played in these modes sounds only slightly less definite and unambiguous than major key music.

The Dorian, Phrygian and Aeolian modes are all very similar to minor keys. In fact, the Aeolian mode is the natural minor scale we met earlier in this chapter. Music performed in these keys sounds rather vague in its punctuation—and, as I said earlier, this can be a pleasant effect for sad or romantic music.

The Locrian mode is not closely related to either our major or minor keys—to our ears it sounds as if a mistake has been made somewhere along the line. For this reason it is only rarely used.

Do different ET keys have different moods?

I would now like to destroy a myth about major and minor keys which is believed by a number of musicians and music lovers. This myth has an excellent pedigree—Beethoven believed in it and lots of other composers and professional musicians have believed it—but it simply isn't true.

The myth is that, even with equal temperament, different keys convey different emotional moods. I'm not talking about the difference between major and minor keys here—they *are* different from each other in the ways I described above. No, the myth says that, for example, E major has a different mood from F major, and D minor has a different mood from B minor. I tested a class of music students on this, and, before we started the tests, three quarters of them believed that different keys had different moods. I then asked them to write down which mood they associated with each key. Without discussing it between themselves, they tended to choose very similar moods for any given key. For example, there was general agreement that A major and E major are "bright and cheerful" and C major is "neutral and pure." If you are one of the people who believes in the moods of keys I suspect that you would also have made the same basic mood choices for these keys. You would probably also agree with the students that E flat major is "romantic and serious." Well, I hate to be a party pooper—but this is all wrong...

Do you remember that committee we discussed back in chapter 1? They met in London in 1939 to decide on the frequencies we were going to use for our notes from then on. They only had to decide on the frequency of one note—because you can then work out the frequencies of all the others from the one you have decided on. After consuming their own weight in chocolate chip cookies, these earnest folks decided that the fundamental frequency of the "A" just above middle C should be 440Hz (440 vibrations per second). They didn't choose this frequency for musical or emotional reasons—they chose it because it was a nice round number which was somewhere in the middle of the range of frequencies being used for the note "A" all over Europe at that time.

Let's go back to look at two keys and their supposed moods:

E major is supposed to be "bright, joyful and lively";
E flat major is supposed to be "romantic and serious."

These two key notes (E and E♭) are next to each other on a piano keyboard—E is only a semitone higher than E♭—yet the supposed moods of their keys are very different. The idea that E major is joyful and E♭ is serious was first proposed by several authors in lists of key–mood links published in the late eighteenth century. The ideas put forward in these lists have survived to this day even though the pitches of the notes involved were not fixed until 1939 and are known to have varied by at least a couple of tones over the years.

On the day of the committee meeting there would have been pianos all over Europe with Es lower than the new standard E♭ and other pianos with E♭s higher than the new standard E. In spite of the large range of frequencies involved, many of the piano owners would have insisted that the key of E on their piano was much brighter sounding than the key of E♭—and they would still insist that this was true after their piano tuner tuned the entire piano higher or lower, to match the new standard pitch. So if there is a key–mood link it can't have anything to do with the frequencies of the notes involved. Also, it can't have anything to do with slightly different tuning systems for the two keys. Piano tuners nowadays are taught to use the equal temperament system, which treats all keys in exactly the same way.

When I was investigating this phenomenon of key–mood links, the only thing left which I could think of which could give a link would be the physical layout of the piano keyboard: the black notes are farther away from your wrists when you are playing, so there might be some subtle difference in the way the black notes are played, and different keys use more or less black notes. On the other hand it could be linked to some effect I hadn't thought of.

I was very skeptical of any key–mood link but thought that the idea deserved to be fairly tested. So, together with a professional musicologist, Dr. Nikki Dibben, I gathered together the aforementioned class of music students and asked them to listen to a specially made tape recording. On the tape were two short pieces of music—one was simple and jolly, and one was dramatic—played four times each in various keys. Between each short piece there was a recording of some Indian sitar music—which was not in any Western key—to prevent the music students from working out how the key of the piano music had changed.

We asked the students (none of whom had perfect pitch and three quarters of whom believed in key–mood association) to try to identify the mood of each piece and the key it was played in. The results showed that the simple, jolly piece stayed simple and jolly no matter what key it was played in, and the students therefore usually guessed that the music was being played in the (supposedly simple) key of G, C or F when in fact it was being played in F sharp major (which is supposed to be complex in nature). We got a similar result for the dramatic music—it stayed dramatic no matter what key it was played in—and the students guessed the keys incorrectly.

Now we have our answer—there is no link between key and mood. I think there are at least a couple of reasons why this myth has become generally accepted:

1. Composers in the past believed it and wrote their music in the "appropriate" keys. This choosing of certain keys for certain moods leads music students to believe the link is real—so they write their own compositions in the "appropriate" keys.
2. When we learn the piano, or any other instrument, we begin with pieces in the key of C because it is the easiest to read as it has no sharps or flats in its key signature (the key signature is written on the left-hand side of each line of music and tells you which notes are to be played sharp or flat throughout the piece). After a few weeks we start to come across music which begins in C but then changes key to G (which has one sharp in its signature) or F (which has one flat in its signature). Changing from C to G makes the music sound brighter, and changing from C to F makes it sound less bright. Because we see that the addition of a sharp to the key signature (as we change to G) adds brightness, and a flat (as we change to F) reduces brightness, we start associating flats with calm or sadness and sharps with brightness. In fact, it isn't true that G is intrinsically bright—it's just that the change up in key makes things sound brighter.

If you change from the key of C to the key of G the brightness of the music increases, but if you move from D to G the brightness decreases: in one case G is the brighter key and in the other situation it's the less bright key.

Now a final note for hardliners—the people who still aren't convinced that there is no link between key and mood...

Earlier in this chapter I mentioned the "truck driver's gear change"—which involves modulating from the key you are in to one a tone or semitone above. The extra "lift" given to any song by this method is very clear—and this is where the argument of mood–key supporters collapses. If you play the relevant section of pop songs which do this on a piano, the "lift" always works, no matter which key you start off in. It works from B♭ (which has 3 flats in it) to B (5 sharps) but it also works just as well from A (3 sharps) to B♭ (3 flats) or from E (4 sharps) to F (1 flat). These last two examples, by the argument of those who believe in key–mood links, should not work because they involve moves from supposedly bright keys to less bright ones. There are other examples of this trick in pop music where the movement upward is more than a whole tone, but once again, it doesn't matter which keys you move from or to — it's the movement upward which infuses the song with new life.

So there we are — it's the movement from key to key which provides a change in mood. The keys themselves don't have moods of their own.

The main things to remember about keys can be summarized in just four short paragraphs.

1. Major keys are a team of seven notes which are strongly related to their team leader. The punctuation of phrases in major key music is generally clear and decisive.
2. Minor keys have a couple of different notes depending on whether or not the tune is going up or down in pitch. The team of notes is not as strongly related as a major key team and the musical experience not as decisive and clear-cut — particularly at the end of phrases. We have learned to associate sadness with this more complex interrelationship of notes.
3. Music changes from major key to major key to keep our levels of interest up, and the same is true from minor key to minor key. Certain changes increase the brightness of the music for a short while, and others can diminish the brightness. The effect does not last long because it is caused by the change itself.

4. Changes from a major key to a minor key, or vice versa, involve a strong change in mood. In this case the major (brighter) or minor (sadder) mood does not fade away—it is not merely linked to the action of changing.

But one thing we haven't covered so far is why generations of unhappy children have been forced at knife-point to practice playing scales on their instruments when they could be having much more fun playing real pieces of music. There are two reasons for this, but I think they are rather feeble when you take into account the degree of boredom which scale practice induces, and the number of kids who have abandoned music because of it. The first reason to practice scales is that it makes you get used to using all the notes on your instrument. The second reason is that lots of tunes involve fragments of scales. In fact, if you look at them closely, most melodies are made up of a combination of arpeggios, repeated notes and parts of scales. In "Baa Baa Black Sheep," for example, the words "sheep have you a..." are four notes of a major scale. Because scale fragments turn up so often in tunes it's helpful to have the whole scale locked into your "muscular memory"* in the same way that it's useful to memorize multiplication tables—it saves a lot of effort later. On the whole, though, I think far too much emphasis is placed on scale practice in the early years of musical education. If the student is going to take up music as more than a hobby, she can move on to scale practice later if she wants or needs to. While I'm feeling all hot-headed and revolutionary, why don't all you music students abandon scale practice in favor of improvisation practice?

Although playing scales is tedious, an understanding of how scales, keys and harmonies work is an important part of musical training. An appreciation of what's going on is very useful whether

* Muscular memory is explained on pages 204–205.

you are playing, improvising, composing or just listening. Next time you are listening to a pop song which suddenly gets a lift in energy, smile knowledgeably, point at the speakers and say "ah... a truck driver's gear change."

Choose your time carefully, though—you are only allowed to be this weird about once a year.

10. I Got Rhythm

I got rhythm, I got tempo, I got meter

There is a strong link between rhythm and vegetarians. I'm not talking about the relative dancing abilities of vegetarians and omnivores here, as I have observed examples of excellent and execrable hoofing from both camps. I mean that there is a strong link between the way we use the words "vegetarian" and "rhythm."

When we say "Kay is a vegetarian" we don't necessarily mean that she eats only vegetables. Lots of vegetarians also eat a wide range of things which are not vegetables, such as eggs, legumes, cheese, and many other foods packed with revolting vitamins. We are using "vegetarian" as a convenient label.

Any piece of music consists of a stream of sounds spread over a certain amount of time, and we use the word *rhythm* to describe how we organize the timing and emphasis of those sounds. But we are using the word as a convenient label. In fact, when we talk of rhythm in this way we are really referring to three things: *tempo*, *meter* and *rhythm*.

The tempo of a piece of music is its pulse rate—how often you would tap your foot to it.

The meter is how often you would emphasize one of the foot taps. For example, if you are listening to a waltz you will emphasize the first tap of groups of three—**one,** two, three, **one,** two, three. If you are listening to rock music (or most other Western music), you will stress the first beat of groups of four—**one,** two, three, four, **one,** two, three, four.

The rhythm is the pattern of long and short notes being used at any particular time. For example, the beginning of Beethoven's Fifth Symphony (Da Da Da Daah) has a rhythm of three short notes followed by a longer one. You can play it as quickly or slowly as you like, but you won't change the rhythm—it will always be three short followed by one long.

Having explained all that, I will now return to the normal, conversational use of the word "rhythm" for the rest of my discussion.

When I was putting this book together I considered all sorts of methods for drawing pictures to explain how rhythms work, but eventually I realized that the clearest and simplest pictures of rhythms were the ones we use in standard written music. The Western system of written music is a diagram which tells musicians which notes to play, when each note should start and stop, and which ones to emphasize. We can use this system for our discussion of rhythm—but don't panic, I am not going to expect you to be able to read music.

Actually, learning to read music is a pain in the neck—and don't let anyone tell you otherwise. It's very interesting for approximately the first ten minutes, when you're learning what it's all about. After that there is a long struggle before you can get your fingers to obey the instructions on the page and produce music. This long struggle is similar to the one involved in learning a language, and the rewards are just as great, but in this chapter we are only going to cover those interesting first ten minutes.

The development of written music

The trouble with ancient history is that, as far as I can tell, the vast majority of it took place a hell of a long time ago. This means that it's very difficult to know when written music was first attempted. There is some evidence of ancient written music of various levels of sophistication in China, Syria and ancient Greece. One of the earliest complete compositions we have discovered so far was inscribed on a tomb about 2,000 years ago. This song, known as the Seikilos Epitaph, is written in an ancient Greek musical notation which tells you which note to sing for each word and how long each note should be. The lyrics of the song encourage us, in the true rock and roll tradition, to shine while we live because life is short.

The written music system we use for Western music dates back to the time when monks and nuns were much more a feature of the national music scene than they are today (although there were also professional musicians around). In the early days, the music would be thought up by a talented musician, monk or nun, and taught directly to his or her friends by singing or playing. Eventually, however, the music became more complicated and composers realized that writing it down would be a handy way of teaching and recording it. Also, by about A.D. 750 the Christian church was beginning to insist that masses should be sung using various standardized rules. For these reasons there was a lot of interest in writing music down. Some early composers used to draw pictures of how the music went up and down but this wasn't very accurate. By about A.D. 800 there was general agreement on the following rules.

1. You need to use different sorts of dots or shapes for different lengths of note.
2. The dots should be drawn one after another, reading from left to right.
3. The dots should be drawn on a "ladder" to show how high or low the notes are.

Writing down music involved giving names to the notes and so they drew their "ladder" and named the notes from A to G. The earliest ladder had eleven lines, like this:

One early idea was to write the notes on an eleven-line "ladder," or "stave," like this, but as you can see, it's very difficult to identify quickly exactly which line or space the notes are on.

This ladder (or *stave* as it is called) had so many levels that it was very difficult to read, so composers split it into two sections, low (or bass) and high (or treble) like this:

The modern stave split into two sections (treble and bass) for easy reading. (Note that middle C is stuck in the middle between the two staves.) The two symbols on the left are just there to identify the treble and the bass sections of the stave because some instruments don't need this enormous range of notes and only use the treble (e.g., violin) or bass (e.g., cello) part for their written music. The symbols are called the treble clef and the bass clef.

Splitting the stave into two parts leaves one note, a C, between the two sections. This is the famous "middle C" you have heard about—it's the middle note on a piano as well. Middle C has no special musical significance—it's just a convenient reference point. Musicians, for example, will say things like "I can't sing this twaddle—my singing range is only up to the G above middle C. Who do you think I am? Freddy bloody Mercury?"

In the illustration below you can see the first lines of two of the songs we have been using for reference throughout this book. I have only used the treble part of the stave because—for these tunes—I don't need the big range in pitch which both staves would give me.

The first lines of "Baa Baa Black Sheep" and "For He's a Jolly Good Fellow."

We will look at how rhythms and notes of different length are noted down in a minute, but first let's look at how the vertical position of the notes tells us how the pitch of a tune goes up and down. If you sing the songs while looking at the written notes you will notice that the notes go up and down the ladder, just as your voice does as the songs progress. I have started both songs on the same note so that you can compare them easily. If you hum these songs you will notice that both of them begin with notes of the same pitch and then there is a big leap up to "black" or a small jump up to "he's," depending on which song you are humming. In every case the musical jump you hear is accurately identified by the vertical position of the notes on the written stave. Remember, as I said in chapter 2, you don't need to start on the correct note ("C" in this case) to sing well—unless you are being accompanied on an instrument. The most important thing is the size of the jumps in the tune. It takes a lot of training to be able to look at a new tune and sing the jumps correctly just by reading the music—but the exact information is all there.

Key signatures

In the last chapter we discussed the fact that different keys involve different notes. For example, the key of A major includes the notes A, B, C#, D, E, F# and G#. If a composer is writing in this key he doesn't want to be bothered with putting a sharp sign in front of every individual F#, C# and G# so he writes a general instruction at the beginning of every line of the music—called the key signature. The key signature of A major, with the instruction to play all the Fs, Cs and Gs as F#, C# and G#, looks like this:

The key signature of the key of A major. The three sharp signs are written on the lines which represent the notes F, C and G (reading left to right). This is a general instruction which means "Every F, C and G should be played as F#, C# and G#."

Writing down rhythms

Different lengths of note

Our two songs involve notes of various lengths and we use different symbols to indicate how long each note is compared to the others. The note symbols and their names were agreed upon centuries ago and are listed below. As you can see, the notes get shorter in a very organized way: we start with a very long note (or "breve") and divide it in half to get a half note (or semibreve), then continue dividing the length of the notes in half to get shorter and shorter notes. Nowadays we also talk about "whole notes" and "quarter notes" and, just to confuse everyone, it has come to be generally accepted that the semibreve, not the breve, should be counted as a whole note—as you can see in the illustration.

Symbol	Name
𝅜	Breve
𝅝	Semibreve (whole note)
𝅗𝅥	Minim (half note)
♩	Crotchet (quarter note)
♪	Quaver (eighth note)
𝅘𝅥𝅯	Semiquaver (sixteenth note)
𝅘𝅥𝅰	Demisemiquaver
𝅘𝅥𝅱	Hemidemisemiquaver (no, I am not joking)

A list of the different symbols for notes of different lengths. Each note is twice as long as the one below it.

If this "halving" system was all we had to describe the length of notes, then our music would be rather dull rhythmically, so we have a couple of additions to give us more flexibility:

1. A dot written immediately after a note means "this note should sound one and a half times as long as normal" (you can see such a dot after the note for "fe" in "fellow"). A double dot after the note is much less common but means "this note should sound for one and three quarters as long as usual."
2. You can write a small "3" above a group of three notes to indicate that "these three notes should take up the same amount of time as two notes would normally." This is a fairly common device and you will have heard it in action many times. Rather less common is the writing of "5" above a group of five notes, or any other similar combination. In every case the message is, "this group of notes should be squeezed into the amount of time allowed for a group of this size minus one," i.e., 5 should only take as long as 4 usually would, 13 should take as long as 12 usually would, etc.

Sometimes notes are written as individuals (this is common for singing) but more often the shorter notes are joined to others to make little groups, as you can see in the songs. This joining together doesn't affect the length of the note—it just helps the musician to read the music.

The shorter notes are usually joined together by their "tails" rather than written individually. The two quavers drawn above (on the left) each have a single tail and would be joined by a single straight line (as on the right of the illustration). Semiquavers have two tails, so they are joined by two lines, and so on.

Looking back to the notes for "Baa Baa Black Sheep," and referring to the list above, you can see that the notes for "have" and "you" are half as long as those for "black" and "sheep," which are, in turn, half as long as the note for "wool." You might have thought that the notes you were singing were just randomly longer or shorter—but you are in fact singing notes whose lengths are closely related to each other.

Stress or emphasis: the use of bar lines

"Baa Baa Black Sheep" is a simple song but it would be simple to the point of dullness if all the notes were the same length and they all had equal emphasis. You will notice that there are vertical lines drawn on the stave at a regular distance apart, which do not have any note or sound associated with them. These are called bar lines. The distance between two bar lines (where we write the notes) is technically called a *measure*—but everyone I've ever met calls it a *bar* so that's the word I'll use.

One of the conventions of written music is that the first note after a bar line is given extra emphasis, or stress. When you sing "Baa Baa Black Sheep" you stress the first "Baa," the word "have" and the first "yes." In the written version of the music, these words appear just after a bar line. If you haven't noticed that you emphasize in this way, try singing the song with deliberate stress on "sheep" and "any." It all sounds a bit Monty Python, doesn't it? Now sing it again with the emphasis where it should be—on "Baa" and "have." You may now be over-emphasizing but the stress is in the right place—just after the bar line—so it sounds OK.

Now, without looking at the written music, sing the first line of "For He's a Jolly Good Fellow" a couple of times. It's only six words, but it involves eight notes: For, he's, a, jo, lly, good, fe, llow. Imagine that you are singing the song in a funny/dramatic way to a friend. For extra effect you have brought a cymbal with you. The noise from a cymbal lasts a long time so you will only hit it once in every line of the song. Sing the song aloud a couple of times and imagine which of the eight sounds you would choose to hit the cymbal on. Will it be "For," "he's," "a," "jo," "lly," "good," "fe" or "llow?" Using my mysterious powers, I can confidently tell you that you will have chosen to make your big crashing noise on either "he's" or "fe." Now have a look at the music—yes, "he's" and "fe" are both immediately after a bar line.

In many cases we would automatically choose the first note of the song as one of the notes to be emphasized. In "Baa Baa Black Sheep," for example, you would have clashed your cymbal on the first word, "Baa," or the word "have." In the case of "For he's . . . ," however, we would not choose the first note "For" because it does not come immediately after a bar line—the song does not start at the beginning of a bar. Just in front of the word "For" there is what musicians call a "rest," a mark (in this case two marks) indicating that the first part of the bar is silent. This may sound a little weird—starting a tune with a silence—but we do it to get all the stresses in the song or tune to fall in the correct place, just after the bar lines. Here are a few examples of songs with the accented syllables printed in bold type.

Tunes which start at the beginning of a bar:

- Twinkle, Twinkle Little Star (**Twin**kle Twinkle **Li**ttle Star . . .)
- Frère Jacques (**Frè**re Jacques **Frè**re Jacques **dor**mez vous . . .)
- London Bridge is Falling Down (**Lon**don Bridge is . . .)

Tunes which don't start at the beginning of a bar:

- When the Saints Come Marching In (Oh, when the **Saints** . . .)
- On Top of Old Smoky (On **top** of Old **Smo**ky . . .)
- Auld Lang Syne (Should **old** acquaintance **be** forgot . . .)
- Greensleeves (**Alas** my love . . .)

If you ever see these tunes written down you will see that the bar lines come immediately before the syllables I have highlighted.

Dividing up the bar—time signatures

You will notice that our two tunes begin with numbers which look like fractions. These numbers (called the *time signature*) tell the musician how many "beats" there are to each bar and, roughly, how long those beats are.

How many beats to the bar? The upper number in the time signature

The upper number in a time signature is the most important one—it gives you the meter of the music, that is, how many beats there are to the bar. Let's take the two most common values for this number: 3 or 4 (4 is by far the most common).

If the time signature has an upper number of 3 then the overall pulse of the music will be: **one** two three, **one** two three—like a waltz. There doesn't have to be a note played for every one of the three beats, and the tune may sometimes involve more than one note per beat; in any case your mind will retain this sense of a pattern of three beats.

Similarly, if the music has a 4 as the top number of the time signature, the music retains an overall "**one** two three four, **one** two three four" rhythmic pulse no matter how many notes are played during each pulse. Most people who have not had any musical training find this difficult to understand. When asked to clap out the rhythmic beat of a familiar tune they tend to clap once per note like this:

Baa	baa	black	sheep	have you an - y	wool?
clap	clap	clap	clap	clap clap clap clap clap	

In this case you get four evenly spaced claps followed by five faster ones.

A musician, asked to clap out the basic rhythm, would do it this way:

Baa	baa	black	sheep	have you an - y		wool?	
clap	clap	clap	clap	clap	clap	clap	clap

Now we get eight equally spaced claps: the musician claps regularly even when there is more than one note per beat ("have you" and "any"), and keeps clapping at the same rate through long notes or silences (in this case there is a one-beat silence between the end of "wool" and the "yes" which follows it).

If we ask the musician to emphasize the note at the beginning of each bar we would get:

Baa	baa	black	sheep	have you an - y		wool?	
clap	clap	clap	clap	**clap**	clap	clap	clap

This tells us that this music has four beats to the bar.

Asking the musician to do the same job for "America," we would hear:

My	coun	try,	**'tis**		of thee	**sweet**	land
clap	clap	clap	**clap**	clap	clap	**clap**	clap

Here we can see that the beats come in groups of three—so the upper number in the time signature will be a three.

The most common time signature (which covers the vast majority of pop music and most classics) divides the bar up into four beats—here is a slightly more complicated example than "Baa Baa Black Sheep":

Oh when the	Saints			Oh when the	Saints	
clap	**clap**	clap	clap	clap	**clap**	clap

Whatever the time signature, words (notes) sometimes occur between the beats ("Oh" and "the" from "When the Saints," and "of" from "America"); sometimes notes last longer than one beat ("tis" and "Saints"); and sometimes a beat happens when there is no note (before "Oh").

If the piece of music has three beats to the bar, then the first beat is the strong one and the other two are weaker (**one** two three, **one** two three). In the case of four beats to the bar, the first beat is the strongest but the third beat (the halfway point in the bar) is the next strongest. The second and fourth beats are both weak by comparison—so the stress in a four-beat-to-the-bar piece is: **one** two *three* four, **one** two *three* four.

Another case of splitting the bar in half occurs if there are six beats to the bar—which is common in Irish jigs. As you can see in the written example, "For He's a Jolly Good Fellow" has six beats to the bar. To get the emphasis correct, this time signature is usually counted aloud in the following way: **one** two three *two* two three, **one** two three *two* two three. The second two (in italics in the middle of the bar) is stressed, but not quite as much as the **one**—for **he's** a *jo*lly good **fe**llow.

Four beats to the bar is, as I mentioned, the most popular case in Western music of any sort and three beats (mostly waltzes), two beats (mostly marches) and six beats (mostly jigs) are used in most of the remaining cases—all other numbers are fairly rare. Modern jazz and modern classical music sometimes make a point of using five, seven, eleven, or more beats to the bar (sometimes just to be clever and unusual), but there are only a few really popular successes:

- five beats to the bar—"Take 5" by the Dave Brubeck Quartet and "Mars" from *The Planets* by Gustav Holst;
- seven beats to the bar—"Money" from *Dark Side of the Moon* by

Pink Floyd and various bits of *The Rite of Spring* by Igor Stravinsky (which also contains many other unusual rhythms);

- nine beats to the bar—used in a particular type of Irish jig, called a slip jig, which divides the bar up into three sets of three beats (**one** two three, *two* two three, *two* two three).

Syncopation

Syncopation is a method of adding an extra layer of interest to music by deliberately emphasizing beats which would normally be unimportant. There are some types of music which deliberately avoid stressing the first beat of the bar in order to give the music a particular feel. Some rock and pop songs keep the "**one** two *three* four" emphasis for the tune but deliberately emphasize beats two and four with the bass guitar and drums, a technique known as "back beat" which was very popular with the Beatles (e.g., "Can't Buy Me Love"). Reggae music takes this idea even further by having the first beat of the bar left almost silent by the rhythm section of the band. In general, the drummer and bass guitar player in a reggae band both emphasize beat number 3 and the rhythm guitarist plays beats 2 and 4.

Back beat, rock and reggae use specific types of syncopation which form part of their identity, but syncopation is used to some extent in nearly all forms of music. You can even syncopate "Baa Baa Black Sheep" to add interest to your performance: "Baa **Baa** Black Sheep Have You **Any** Wool..."

Syncopation doesn't even need to involve a full beat of the music. You can emphasize just a part of a beat—"Have you a**ny** wool?" Wherever you come across it, syncopation makes the music sound less predictable and more sophisticated.

What length are the beats? The lower number of the time signature

The lower number of the time signature is always a 2, 4, 8, 16 or 32. Of these, 4 is by far the most common and, really, could be used in all cases. The reason for this is that, although the choice of lower number changes how the music looks on the printed page, it doesn't have any real effect on how the music sounds. This odd fact needs some explanation—so I'd better get on with it.

I am going to use the term "whole note" rather than "semibreve" in the following discussion to keep things as clear as possible.

When a composer writes a "3" over an "8" as the time signature, she is saying that each bar will have three beats, each of which is one eighth of a whole note long—so "3" over "8" simply means "three eighths of the length of a whole note per bar, please." Similarly, 5 over 4 means that each bar will contain five beats which are all one quarter of a whole note long ("five quarters of the length of a whole note per bar, please").

The problem is that no one has ever said how long a whole note is. In the most common time signature (4 over 4), each bar is one whole note long (four quarters)—but if you pick twenty different pieces of music and measure how long the bars are with a stop-watch you will have twenty different results. Although no one knows how long a whole note is, we know the approximate range involved—a whole note (equal to a bar of 4/4 time) will generally be shorter than six seconds but longer than one second. This is, of course, an enormous range—so musicians need more information than just the time signature if they are to play the music as the composer intended.

Musicians get a bit of help from a word printed above the beginning of the music which tells you (usually in Italian or German) how fast the music should be played—"presto" means fast and "adagio" means slow. But these are, of course, rather vague terms—two different performers might differ in speed by 50 percent or more.

In an attempt to minimize vagueness, composers often put metronome marks next to the speed-indicating word. They will, for example, draw a note of a certain length (e.g., ♪) followed by an "=" sign followed by a number (e.g., 120). The number tells the musician how many of that type of note would last for one minute, so "♪= 120" means that you could fit 120 of these eighth notes into one minute—or that each ♪ note lasts for half a second. (A metronome is just a device which you set to tick as slowly or as quickly as you want it to—in this case you would, of course, set it to tick 120 times per minute and then play along in time with the ticks.)

The use of metronome marks sounds logical until you find out that most professional musicians don't take much notice of them—they just play as fast or slowly as they want to. For example, I have a couple of recordings of the famous guitar concerto by Rodrigo in front of me (the Concierto de Aranjuez) and the CD boxes tell me that John Williams plays the romantic second movement in just under ten minutes but Pepe Romero completes the same piece in just over twelve minutes (that's a 20 percent difference, and it has nothing to do with how well they both play their instruments; Pepe just likes the extra romance of playing it more slowly). The piece has a metronome marking at the beginning but also includes vague instructions to speed up and slow down at various points—so there is no "correct" completion time. It is also well known that many composers cannot decide on the correct number for the metronome mark and, if they have been recorded, often go considerably faster or slower than their own instructions indicate.

So there we have it—the bottom number of the time signature gives us little or no information about how the music sounds. You could, for example, write a piece in 3 over 8 and mark it "slow" or write it in 3 over 4 and mark it "fast"—the music would sound the same. In fact, the music in 3 over 4 might be played faster than the music in 3 over 8 because "fast" and "slow" are such vague terms. No matter how well trained you are you cannot tell whether the modern jazz piece you are listening to is in 3/4 or 3/8 (or tell the difference between 5/4 and 5/16). If you had to make a bet, your only guide would be that, traditionally, bigger numbers on the bottom of the time signature are usually associated with faster music. For example, a romantic Viennese waltz will almost certainly be written with a time signature of 3/4 rather than 3/8, 3/16 or 3/2. Similarly, an Irish jig will generally be written in 6/8 rather than 6/4 or 6/16.

This quick guide to how written music works can be reduced down to five points.

- The vertical position of the notes tells you the size of the jumps in the tune.
- Notes of different lengths have different symbols.
- Notes which appear immediately after the bar line are (usually) emphasized or stressed.
- The top number of the time signature tells you how many beats to the bar.
- The length of the beats is given approximately by the lower number of the time signature together with a word indicating the speed of the piece. Sometimes this speed information is given more accurately in the form of a metronome mark.

Dancing and rhythm

If we go back to my point that rhythm can be divided up into rhythm, tempo and meter, it may come as a surprise that rhythm is the least important of these three components where dancing is concerned. The lengths of the different notes in the music are far less important to a dancer than tempo and meter. The tempo of the music tells you how fast to dance and the meter tells you what type of dancing you are involved in. The vast majority of Western modern dance music is in a meter of four beats to the bar with a simple 1, 2, 3, 4 count and so the only important variable is the tempo.

There is a widely held misconception that your heart rate tries to match the pulse rate of the music you are listening to. This belief probably arose because the range of pulse rates for music and human beings is similar. The tempo of music is usually between 40 and 160 beats per minute (bpm), and human pulse rates range from about 60 for a relaxed person with a slower than average heart, up to above 150 bpm for a healthy young adult dancing his socks off.

If you are a healthy young adult dancing in a club, the music will have a tempo between 90 and 140 bpm, and your heart will be batting along at a similar rate. But the exciting noise of the music is not what's driving your heart rate up—it's the dancing. The music will have an average beat rate of approximately 120 bpm because this is a sustainable, fun rate at which to be waggling your various bits about. You can move your body and limbs twice a second (120 bpm) for an hour or so without anything snapping off. Your heart will be racing along, faster than normal because you are consuming a lot of energy. Next time you are in a club, try taking the pulse of a friend who is dancing and comparing it to one who is slumped at the bar. The dancer will have a heart rate close to the pulse of the music, but the slumper's heart will be much slower. But don't forget—we slumpers need love too.

Of course, dancing is nothing new, and people have always enjoyed shaking their bits in the general direction of people they fancy. Dancing usually involves raising your heart rate and the waltz does this in two ways. The first reason for an increased pulse is the physical effort involved. A waltz has a tempo of approximately 100 bpm. This is a lower tempo than the 120 of the music found in the clubs today, probably because you have more work to do. Two people have to coordinate their movements and then steer themselves around the dance floor rather than merely hopping up and down in one spot. It's also fairly certain that sweating like a pig used to be less fashionable than it presently is.

The second heart rate accelerator of the waltz is the reason why it was almost banned when it was first introduced into polite society—it was a method of actually getting your hands on the body of your beloved, or be-fancied. Even the torrid waltz, however, is not quite as "hands on" as a much earlier dance with a three-beat meter, the volta, which was a favorite of Queen Elizabeth I. The volta involves a lot of lifting and lowering of the woman, with plenty of opportunity for mutually enjoyable, accidental hand slippage. If enough of you readers vote in favor, we could try to get it reinstated as the biggest dance craze around, as it was in London in the late 1500s.

Rhythm and polyrhythm

If you are sitting and listening to music, rather than dancing, you have a much deeper appreciation of the subtleties of rhythm. Meter, rhythm and tempo all play their part in our enjoyment. Although our pulse rates do not link themselves to the tempo, we certainly find slower tempos more relaxing and faster ones more exciting. This tension is probably linked to our dislike of uncertainty, particularly our fear of not being able to cope with a situation. If the sounds we are hearing are moving at walking pace or slower, then there is no need for anxiety. If things are rushing along we might need to be prepared to run away or protect ourselves.

One good way of getting a lot of emotion out of a piece of music is through a technique called *rubato*. The word means "robbed." The musician steals a bit of time off a couple of notes in order to make the note before or after them last a little longer. Instead of hearing the notes with a steady rhythm, "Daa, Daa, Daa, Daa, Daa, Daa," we get the effect of a rush up to a longer note, "Daa, Daa, Daa, Da-Da-Daaaa," which adds drama and romance to the music.

Changes of meter are a good way of keeping our interest levels high, and unusual meters such as seven beats to the bar also keep us interested because they sound unsettled or incomplete. But a lot of Western music is not very rhythmically adventurous or sophisticated. The music tends to concentrate on four or three beats to the bar with simple, regular sub-divisions of these beats. African and Asian musical traditions and several others often employ greater rhythmic complexity, including the use of polyrhythms.

Polyrhythm involves playing two or more non-collaborative rhythms at once. To explain what this means, let's take the example of you and me tapping out rhythms on your dining-room table. If we both tap together in groups of four, we are just tapping the same rhythm. If I tap eight times to every four of yours, then our rhythms will collaborate—we will be hitting the table at the same time at regular, frequent intervals, and every extra tap of mine will fit exactly into the gaps between yours. If, on the other hand, I tap the table five times for every twelve of yours, then our taps will not coincide very often, and most of the time my taps will not be synchronized with yours in any obvious way. The two rhythms are no longer collaborating. We are tapping polyrhythmically.

The idea of polyrhythms is not completely new to Western music. Mozart used a pulsing accompaniment based on threes against a tune based on twos in the second movement of his Piano Concerto No. 21 (this piece is nowadays known as "Elvira Madigan" because it was used in a film of that name). More recently, some jazz and rock bands have used polyrhythms and I think that they will gradually become more commonplace. This will be useful for people like me, because when people tell me I'm dancing really badly, I can just claim to be dancing to the other half of the polyrhythm.

11. Making Music

The myth of musicality

For some reason, a lot of people think that if you haven't studied a musical instrument by the time you are twenty then it is already too late. Also, people who didn't study music as children, or who had a horrible time learning an instrument when they were kids, often declare themselves to be "unmusical" but they would "love to be able to play an instrument." If you ask such people about any other skill they would like to acquire, such as making pots or knitting, they don't declare themselves to be "unpotterly" or "unknitty," they quite sensibly say that they probably could do it if they bought the equipment and took some lessons. They realize that they would probably never be able to compete with professionals but could eventually produce worthwhile stuff and have fun in the process.

There is general agreement that anyone can acquire almost any skill to some level of competence. But music is considered a special case—apparently you're either musically talented or you're not. Thankfully this view is entirely wrong: playing a musical instrument is just a skill to be learned like any other. Some people (especially children) pick up the skills involved faster than others (which is true of any skill), but everyone gets better with time and effort.

Another myth about music is that it takes years to learn an instrument. This is only true if you have very high expectations. If you want to play Beethoven sonatas in public, then yes, it will take more than ten years and you will have to practice for more than an hour a day. If, on the other hand, you want to play a Bob Dylan song at a campfire singalong you could probably be ready in a month if you practiced for a few minutes on most days. By the end of the year you could have more than a dozen songs which you could play. It is also very important to remember that learning an instrument is a lot of fun from the beginning. The only pain in the neck is that it involves a lot of repetition—but even that is OK when you can hear yourself getting better and better.

One of the most daunting things about musicians is the way they seem to be able to remember an inhuman amount of notes and regurgitate them at will. This is particularly true of musicians playing classical music from memory: sometimes the musician has to produce thousands of notes in exactly the correct order and if they get even one wrong it will be noticed by the audience. If a non-musician sees this sort of feat it puts her off the idea of learning an instrument because she is sure that her memory (and fingers) couldn't work that well.

Without in any way diminishing the achievement of such performers, it is useful to know that they are being assisted by something known as "muscular memory." Obviously muscles can't actually remember things, but complex sequences of muscular movement can be stored by the brain as a single memory. If this sounds a little unlikely, just think about how little mental effort and memory you need in order to tie your shoelaces every morning. Next time you tie your laces just watch your fingers—it's an amazingly complicated set of muscular movements, but your brain just sends out a single instruction, "tie shoelaces now." A trained musician can render a whole piece of music down into a sequence of linked "shoelace-tying" sets of instructions. The brain is not sending out instructions for each finger movement; it's saying "next comes the bit with the jiggle in the middle; now there's the bit with the three loud chords." Getting your brain to do this for a piece of music requires a lot of repetition or practice—but there is nothing magical about it. The magic is in the sounds we create and how people respond to them.

So those of you who have been saying "I'd love to play a musical instrument but I'm just not musical" can go down to the music shop on Saturday and buy an instrument. Everyone is "musical"—becoming a musician is simply a matter of learning a skill. You will be worse than some and better than others but you will be a musician.

If you have decided to take the plunge, the following notes might help you choose the most appropriate instrument—all instruments involve a learning process but some are kinder to the beginner (and their neighbors) than others.

Choosing an instrument

There are far too many instruments in the world for me to list them all here, but I can give you some pointers about some of the more common ones. Musical instruments can be categorized in a number of different ways. For example, there are instruments which can only produce one note at a time (such as flutes), ones on which it is difficult to produce more than one note at a time (such as violins) and those on which it is easy to produce lots of notes at once (such as pianos).

Another categorization which may be useful to a beginner is that some instruments have definite places where you put your fingers to get a certain note (e.g., pianos, flutes, guitars), and some instruments don't (e.g., violins and trombones). For those of you who have never held a violin or trombone, this may need some explanation.

To take the simplest case of an instrument with set places for your fingers for each note, look at the piano. In the illustration below you can see that if I want to play the note we call "middle C" I just have to press the correct key. Pressing that key will always produce that note and I can't get that note by pressing any other key. Because the key itself is quite large compared to my fingertip, I don't have to be very accurate in hitting it, so long as I miss the keys on either side of it. (By the way, I'm assuming that the piano has been tuned.)

To hear the note "middle C" on a piano, all I have to do is press the correct key with one finger.

Now let's have a look at how you get notes from a guitar. In the illustration opposite, you can see that I have shortened the string I am about to pluck by pressing one of my fingers on the neck of the guitar behind one of the frets. When I pluck the string, I will get the note relevant to the length of the string between the fret and the bridge of the guitar. Once again, I don't have to be extremely accurate—my finger can be right up against the fret (as in the left-hand photo below) or a few millimeters away from it (as in the middle photo). The note will be the same in both cases because the position of the fret determines the note produced, not the exact position of my finger.

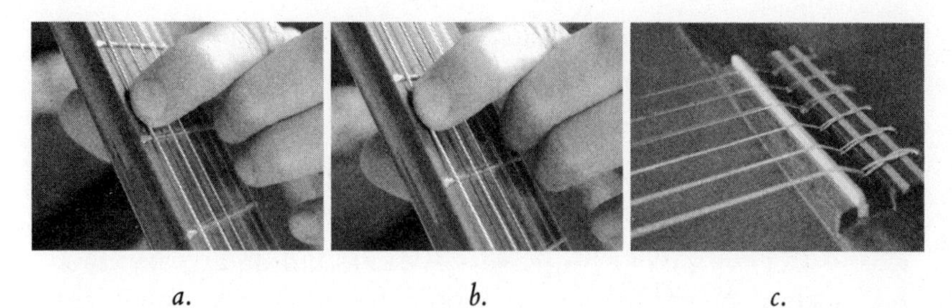

a. *b.* *c.*

To get a chosen note on a guitar I need two fingers. One finger plucks the string and a finger from the other hand presses the string against the neck of the guitar to trap it over one of the frets. As is the case with the piano, my fingers only need to be accurate within a few millimeters. The finger positions shown in the two illustrations produce the same note because—even though my finger has moved—the fret stays in the same place. The final photo shows the bridge of the guitar, which holds the other end of the strings.

If we now look at a violin, you can see that there are no frets—the pitch of the note produced is determined by the exact position of your finger as it traps the string against the neck and makes it shorter. Wherever you put your finger, you get a note—but only a very small percentage of those notes will be the ones you want. In this case the position of your finger has to be accurate within a millimeter or so.

The lack of frets on a violin neck means that it is much more difficult for a beginner to identify where to put their finger down to shorten the string. Also, the position of the finger must be accurate to within a millimeter or so.

If you want to play the first three notes of "Three Blind Mice" on a piano, you use just one finger to ding out "E, D, C"—it takes about half a minute to learn and you can sound proficient in two minutes. To play the same notes on a guitar involves both hands (one for plucking, one for the frets). It takes a little longer to learn and to become proficient, because frets are not as easy to use as piano keys. Nevertheless you should be making a convincing "Three Blind Mice" noise after about twenty minutes: you learn which frets you have to get your fingers behind for each note and you only have to be accurate in your finger placement to the nearest few millimeters.

The situation is entirely different if you now try to play the tune on a violin, even if we ignore the fact that using the violin bow is pretty difficult in the first place. Let's imagine that you have had weeks of training on the use of the bow, but this is the first time you have tried shortening the strings by pressing them against the neck of the instrument. There are no frets and there is nothing to use as a visual guide, so it's very difficult to know where to put down your fingers. If you press your finger down even two millimeters away from the correct place you will produce a note which is noticeably wrong. Also, because you have to hold the instrument up under your chin, you're looking along the neck and it's difficult to judge these distances properly.

Taking all this into account, I think it's perfectly reasonable to declare that the guitar and piano are kinder to total beginners than the violin. However, don't be misled into thinking that a trained violinist is more skilled than a trained pianist or guitarist—nearly all instruments require similar amounts of skill once you have progressed beyond the beginner stage. "Why," you might well ask, "if the violin is more difficult to start with, doesn't it remain more difficult?" The answer to this question is that musical training is designed to get the best out of every instrument. Within a few weeks a trainee pianist will be asked to play more than one note at once most of the time. By the time he has been learning a couple of years he may be playing four, five or even more notes at the same time. Violinists will only rarely be asked to play more than one note at a time and will not attempt even two together until they have been training for several years, because it is much more difficult to do on the violin. In each case the musician is trained within the limits of the instrument she is trying to master.

Which instrument should you take up? Well, of course, that's up to you. I wish you luck whichever one you choose. My only advice is that, if you are over twenty and you've never played an instrument before, don't put yourself off by starting with one of the instruments which are really tough on total beginners—the slide trombone, French horn, bassoon, violin, viola and cello fall into this category.

Slide trombones, for example, make a marvelous noise but they involve more skill than most instruments for the beginner. To start with, you have to learn how to use the sliding bent tube which makes the instrument longer or shorter—a longer tube gives a lower fundamental frequency. There are seven different correct positions for this slide but no indication of where they are—you just have to practice getting it right. Once you have pushed the slide into the correct position, you can get one of about ten possible notes depending on the type of farting noise you make with your lips—which is determined by how tightly you press your lips together and how hard you blow. These different notes are the harmonics of the tube length you choose by moving the slide. In the early stages of learning it is very easy to get completely the wrong note, either by putting the slide in the wrong place, or by blowing too hard or pursing your lips in the wrong way. I have a lot of admiration for people who learn one of the instruments which are extra tricky for beginners, but my admiration is tempered by NIMBYism—I wouldn't want a trainee trombonist living next door.

If you want to take up an instrument you blow into, I suggest starting with the flute, saxophone or trumpet, instruments with clear positions for your fingers. You can then change instruments after a few months if you want to. You might also want to consider portability and storage (it's easier to store and carry a clarinet than a harp), and how satisfying the instrument is to play solo—because you will be practicing on your own most of the time. For example, there is a far more interesting repertoire of printed music for the solo piano than there is for any other instrument. By the way, if you do decide to go for the piano, I recommend one of the high-quality electronic keyboards rather than a real piano, because you can practice with headphones on without disturbing your neighbors, and you can fool around with all the other sounds they make if you get bored practicing.

Finally, I would suggest that you join an evening class or find a teacher. You can make decent progress if you have a half-hour lesson once a week or so and then practice for an hour or so every week.

How do composers learn to compose?

To most non-musicians the process of composition is totally mysterious. Other jobs or hobbies seem pretty straightforward by comparison. If you spend a long time training, you can become a dentist, portrait painter, crane driver or gardener—and the sorts of things you would have to study have very little mystery to them.

Training to compose music of any sort always involves a lot of trial and error. If you are starting a rock band with your friends you usually start off playing other people's music, but eventually you might start to write your own. This can either be a collaborative thing or might involve just one writer. Initially the music will be very imitative of your favorite musicians, but eventually your own musical personality will come through. This type of informal "on-the-job" training for writing music is very common for rock and pop musicians.

It is also possible to study composition at a college or university. At one point I studied composition, and my friends used to ask me what I did with my tutor—apart from drink tea. I suppose most of them thought that I would think up a tune and then my tutor and I would mess about with it until it became a piece of music. Nobody really understood how you could train to do something as "artistic" as composition. Eventually, to explain what was actually going on, I developed the following analogy.

Imagine you are training to be a TV comedy scriptwriter—you are a fairly amusing person and you have developed some writing skills. You have thought up a funny situation but it just doesn't work very well once you have written it down. If you take the sketch to your tutor, she will use her experience to try to find ways of getting the maximum amount of audience pleasure out of your original idea. She will probably suggest changes, such as:

- maybe it's too long or too short...
- maybe all the interesting stuff happens too early or too late...
- maybe there are too many or too few people involved...
- maybe you need to remove a line which gives the punch line away too early—or add one to make the joke clearer...

The tutor (if she is any good) will not work directly on the piece with you—but she will suggest specific areas you might want to look at.

This is basically how composition is taught—you go in with a musical idea and the tutor assists you with suggestions like the ones above and also helps with the technical issues. In music there are lots of technical issues. If you are writing for instruments you don't play yourself you need to learn a lot about them if the music is to be playable. There are obvious simple mistakes—"flutes can't go that low"—and less obvious ones—"this trumpeter's lips went completely numb about two minutes ago."

So the sort of training you get is pretty straightforward, and a lot of it is similar to the sort of training you would get as a scriptwriter—it's all about content and timing.

As far as the original musical idea is concerned, that's no great mystery either. You don't necessarily start with a tune—it might be a rhythm or a "bass line" (a repetitive bass accompaniment). It might be something you found yourself humming or something you misplayed on the piano. For example, the English composer Vaughan Williams based the second movement of his Third Symphony on a single mistake he heard an army bugler make while he was working as an ambulance driver during the First World War.

Anyone can make up a tune. Just hum to yourself for a couple of minutes and eventually something worthwhile will pop out. Or you could sit at the piano playing with one finger quite slowly—occasionally tunes will appear. The difficulty is not in producing tunes. It's in remembering them and then developing harmonies and, finally, writing it all down or recording it. This is where musical training is useful; it will help you to remember your tune and write it down. Unless you can record your idea or write it down, it will be difficult to work on and improve it and, in any case, nobody else will ever be able to hear it because you will probably forget what you wrote.

So whether you are writing an opera or the next rock classic, the basics of musical composition go like this:

1. Come up with one or two (generally very short) musical ideas.
2. Write them down or record them (there are computer packages to help you).

3. Use them to develop accompanying music (i.e., if you started with a tune, try different accompaniments; if you started with a bass line, develop a tune and/or chords).
4. Write it all down or record it.
5. Now organize the overall timing. Just like a well-told joke, does it need to be made longer by repeating bits? Does it need thirty seconds of droning, mysterious introductory music?
6. Write it all down or record it.

And there you are—your own composition. At first you might only produce the musical equivalent of "knock knock" jokes, but eventually your stuff will become more sophisticated and (hopefully) enjoyable to others. (If you make a fortune from following this advice, please send a check for 5 percent of your annual income made out to "John Powell"—credit cards are also acceptable.)

While we are on the subject of composition, I would like to offer you a few notes on a subject which baffles a lot of people when they first encounter classical music.

Why do classical pieces have such long, complicated names?

If you listen to any classical music station for an hour or so you are bound to hear the announcer say something like, "That was the allegro first movement from Mozart's Concerto for Piano and Orchestra No. 17 in G major, K453" or, "Next, we are going to hear Prokofiev's Piano Concerto No. 3 in C major, Opus 26." This is one of the reasons that many people find classical music unapproachable—even if they hear something they like, it's difficult to know how to ask for it in a shop. The pieces need to have names if we are going to buy a recording or discuss them—but why do the names need to be so complicated?

One way to disentangle the origins of the names is to explain a few examples. Let's start with the Mozart one above.

Decoding "The allegro first movement from Mozart's Concerto for Piano and Orchestra No. 17 in G major, K453"

CONCERTO FOR PIANO AND ORCHESTRA

The two most common types of classical music which involve an orchestra are the symphony and the concerto.

A full orchestra playing a symphony involves about a hundred musicians but generally they don't all play at the same time. The composer chooses which instruments do what at any given moment. If the music is to sound wistful, the composer might write a tune for a single oboe, accompanied by violins and a harp. The same tune might be reintroduced later in a more dramatic section played by brass instruments, accompanied by drums. These changes in instrumental tone help to sustain the listener's interest. The performance of a symphony can therefore be described as an orchestra all working together as a team—passing the musical work around and occasionally all playing together.

The only difference between a symphony and a concerto is that a concerto also involves a soloist who sits or stands at the front of the stage and shows off throughout the piece. The soloist might be playing any instrument (in this case it's a piano but there are also concertos written for cello, guitar, trumpet, etc.), but the point of the solo instrument is that it adds extra drama to the music. The soloist does more work than any other musician, as he rarely takes a break (he also gets paid more). For a concerto the composer might write music for strings and soloist together; followed by brass and soloist; soloist alone; full orchestra alone; then full orchestra with soloist, and so on. Musical "conversations" or even "arguments" can be set up where the soloist is playing one thing while the orchestra is responding with something else. Basically the on/off relationship between the soloist and the orchestra makes the music even more varied than a symphony—and also more interesting to watch at a concert because you have a "star of the show." This description only fits concertos written since around 1800. Before that time the word "concerto" just meant "a piece of music," which might involve a soloist (as in Bach's Concerto in A for Violin and Orchestra), or might not (as in Bach's Brandenburg Concertos).

When we discuss a particular concerto, we usually refer to the solo instrument as well as the orchestra, so we get names like "Concerto for Piano and Orchestra."

FIRST MOVEMENT

Classical music enthusiasts might find the following prosaic description of the organizational motives of composers very difficult to accept, but it is important to remember that composers have always needed to have a professional attitude towards their music. Swanning around, insisting that you are an "artist," won't pay the rent.

Traditionally, classical music was written for live concerts, a straightforward entertainment which resulted in the composer and the musicians getting paid and everyone else having a good time for an hour or two. From the point of view of a working composer like Mozart, the following two guidelines are important:

1. The orchestra should change pieces every few minutes to keep interest levels up—so the audience doesn't start chatting, nodding off, or playing tic-tac-toe.
2. The pieces should be presented in groups of three or four to cut down on the amount of disruption and effort involved in applauding.

From these two fairly simple rules, most composers from about 1750 on have presented three or four individual pieces of music (each between five and twenty minutes long) as parts (or *movements*) of a bigger piece which they call a symphony, concerto, or in the case of solo instruments, a sonata. There is a short silence between each movement during which you are not supposed to clap—yet another source of confusion to the newcomer. You are only supposed to applaud at the very end.

In some cases there might be a musical link between the movements, but they can also be specifically designed to clash (in order to keep interest levels up). In the piece we are discussing there are three movements: the first is about thirteen minutes long; the second is about ten minutes; and the third is about eight minutes.

ALLEGRO

Apart from having different tunes, the movements are often pieces with different speeds: it is common (but not a rule) to start with a fast movement, then have a slow romantic movement, and then finish with another fast movement. The movements can be referred to either by their number (first, second, etc.) or by their speed, which is usually indicated in Italian, French or German. In this case the word "allegro" is simply the Italian for "fast."

Having given us the number of the movement, our radio announcers do not need to tell us the speed as well—but they often do.

MOZART

Obviously we need to know the name of the composer if we are to track the piece down.

NO. 17

Mozart wrote over twenty piano concertos, so we need to know which number it is. The concertos are numbered in the order in which he wrote them.

IN G MAJOR

This is a totally pointless piece of information unless Mozart only used G major for one of his piano concertos—in which case this information could be used instead of the number to identify it. Apart from this minor point I don't know why everyone involved in classical music broadcasting keeps telling us what key things were written in—it makes no difference to any of us.

K453

A music historian called Köchel (pronounced "Kerkul") spent a large part of his life cataloging Mozart's music and numbered every piece in the order it was composed. So now we refer to each piece by its "K" or "Köchel" number (as well as the separate concerto number).

COULD WE SHORTEN THE NAME OF THIS PIECE?

Yes, we could: the radio announcer could have given us all the information we need by calling it any of the following:

- "The allegro from Mozart's Concerto for Piano and Orchestra No. 17"
- "The first movement from Mozart's Concerto for Piano and Orchestra No. 17"
- "The allegro from Mozart's Concerto for Piano and Orchestra, K453"
- "The first movement from Mozart's Concerto for Piano and Orchestra, K453"

Let's have a look at a few more examples.

Prokofiev's Piano Concerto No. 3 in C major, Opus 26

This is very similar to the title of the Mozart piece, except for the word "Opus," which means "piece of work." In music the opus number generally refers to a piece of published work—such works are numbered chronologically—so this piano concerto was the twenty-sixth work which Prokofiev managed to get published (only very high-quality pieces get published).

Tchaikovsky's Symphony No. 6, Opus 74, the "Pathétique"

As I said earlier, a symphony involves an orchestra without a soloist—although the composer might choose to use individuals for short solos during the piece. Most symphonies have four movements, each of which is usually between five and twenty minutes long. Symphonies are numbered and sometimes also have names—as in this case, the "Pathétique," or "sad."

The Prelude, Fugue and Sarabande from Bach's Lute Suite No. 2 in C minor, BWV997

Bach, and other composers of his time, often grouped together about six short pieces into a *suite*. These suites usually began with a *prelude* ("pre" meaning "before," and "lude" meaning "play") and then were followed by several dances. The fact that they were called dances merely meant that they had the distinctive rhythm of certain dances (in the same way that a composer might call a movement of their symphony a waltz because it has an "um-pa-pa" rhythm), but you were not supposed to dance to them. The dances in Bach's suites had names such as *sarabande*, *gigue* and *minuet*. The rhythm of a sarabande is like a slow waltz.

The *fugue* in this title is a word which means "flight" in English. As far as music is concerned, a fugue is generally a difficult piece which involves lots of counterpoint—the playing of more than one tune at the same time.

This particular suite was written for a single musician playing the *lute*, which is similar to a guitar. The BWV number is, like the K number for Mozart's work, a catalog number to identify the piece accurately.

Beethoven's Piano Sonata No. 14 in C sharp minor, Opus 27, "Moonlight"

Sonatas are almost always pieces for one or two instruments and are generally in three or four movements (for the usual reasons—see my earlier comments on "First Movement"). A piano sonata is always written for a solo piano but a violin or cello sonata will usually include a piano accompaniment. (There is a rather stingy tradition among composers, concert promoters, radio announcers and CD sleeve designers to relegate the pianist to the rank of "accompanist" rather than treating him like one half of a duet—which would be closer to the truth.) By the way, Beethoven didn't call this piece "Moonlight"—he called it "Sonata Quasi una Fantasia." One of the critics who reviewed the piece, a man called Ludwig Rellstab, wrote that the first movement reminded him of the moonlight on Lake Lucerne—and the idea stuck.

Now that you know all about the opus numbering, key naming, and so on, we can deal with a few more bits of classical music jargon.

STRING QUARTET

A string quartet is written for, and performed by, two violins, a viola and a cello. There are usually four movements.

STRING TRIO

A string quartet without the second violin.

STRING QUINTET

A string quartet with an extra viola or cello.

PIANO QUINTET

A string quartet with a piano.

CLARINET QUINTET

A string quartet with a clarinet.

CANTATA

A piece for choir and (usually) orchestra with occasional solo singers. These are often quite long (an hour or so) and made up of lots of five- or ten-minute movements.

CHAMBER MUSIC

The original idea for chamber music was that it should be performed by a small number of musicians in a room (*chambre*) rather than a concert hall. Nowadays the term just means any music written for up to about ten musicians (such as string quartets or quintets).

LIEDER

"Lieder" is the German word for "song." The term "lieder" generally describes a solo singer (with clasped hands and a fancy dress, or clasped hands and a bow tie) singing to a piano accompaniment.

Now we have decoded the titles of classical pieces, I would like to stay on the subject of classical music for a little while, in order to answer another puzzling conundrum.

How do conductors justify their enormous paychecks?

If you go to a classical symphony concert you will find that there are about a hundred people on the stage doing all the work while one person, with their back to you, wafts a stick around. Surprisingly, it is the stick-waggler who is the star of the show—and the best paid member of the band. To many this arrangement seems unfair. Apart from the fact that waggling a stick is demonstrably easier than playing an instrument, no one in the orchestra seems to be paying a blind bit of notice to the waggles of the stick in question.

In fact, by the time everyone gets on stage, the conductor has done most of her work. This preliminary work takes place during rehearsals and involves making a lot of choices about speed, balance and loudness. "Why do these choices need to be made?" you might ask. "Didn't the composer make all these choices in the first place?" Well, the surprising answer to that question is "No." The amount of information the composer writes down on the printed page varies according to the composer and the historical date of the composition: for example, music written before 1800 often has no indication of the speed at which it should be played, or any other extra information—all you get on the page is a stream of notes.

Music composed in the past 200 years usually has information written alongside the notes telling the musicians when to speed up or get louder, but these instructions tend to be rather vague. A piece of music will commonly state what speed to start at (by means of a metronome mark, which tells you how many notes per minute you should play), and might then tell you to slow down for a while—but will not generally indicate how much to slow down. Similarly, there will be various indications on the written music as to when to play louder—but only a rough indication about how loud to get.

You might wonder why composers aren't more exact in these matters, but the point is that a single page of orchestral music already contains hundreds of pieces of musical information, as you can see in the example below, and more detail might cloud the musical message rather than clarify it. In any case, a bit of variability adds interest to each new performance. One of my favorite stories about this flexibility concerns the Finnish composer Sibelius. He was listening to a rehearsal of his violin concerto when the violin soloist asked him a question about how to interpret a certain passage. "Do you prefer this passage played like this... [he then played it with a very sweet tone], or like this... [playing it with a dryer tone]?" Sibelius thought for a few seconds and then pronounced judgment: "I prefer both versions."

Apart from all this vagueness about loudness and speed, there is the choice of the overall balance of the sound of the orchestra, which will, of course, change during the piece. Quite often the written music gives no indication that, for example, the violins need to get gradually louder during a particular section because they will be taking over the tune from the woodwind instruments in the next bit. The conductor can color and shade the overall sound of the piece by deciding which instruments should play loudest at any point. This is not as straightforward as it might seem because you don't always want to do the obvious thing of playing the tune loudly over a quieter background harmony.

There are, in fact, hundreds of important decisions to make during rehearsals-and it's the conductor's job to make them all. All conductors make different, equally valid decisions and some of them change their minds about how a piece should be played as they get older. This means that each concert or recording of a piece of classical music is unique, which is why people often have several recordings of the same piece.

A typical page from an orchestral piece, showing that each page contains hundreds of pieces of information. In this case, all the information represents about twelve seconds of music for twenty-seven different types of instrument, all playing at the same time (e.g., the top line is for a flute and the bottom line is for the double basses).

Things are better organized nowadays, but toward the end of the nineteenth century it was difficult for the conductor to pass on his ideas about the music to the orchestra, because many members of the orchestra weren't actually there for rehearsals. Hangovers, illicit love affairs and better paid gigs meant that a lot of the orchestral members had better things to do in the afternoons than go to rehearsals, so they paid "deputies" to take their places. The deputy would play the instrument of the missing musician to make sure there were enough violins, clarinets, or whatever, at the rehearsal. This was fine if only a few people did it, but it eventually became fairly common for over half the orchestra to send deputies. Sometimes the deputy would also play the concert, but generally a large number of the players at the concert had not been at the rehearsal.

Another problem conductors had to deal with in those days was naughty percussionists. In a lot of orchestral scores you need two or three percussionists to do a lot of impressive banging and crashing at particular climaxes of the music, but sometimes they have nothing to do for twenty minutes or more. Obviously, these long breaks bring out the natural tendency of the musician to slink off to the pub for a quick pint. Back in 1896 Sir Henry Wood (the conductor who invented the Proms) found that the only way to stop the slinking percussionists was to lock the exit doors. In his autobiography he describes the result: "I would see these fellows creeping up, one by one, bending double in the hope of being hidden by the music stands. They would gently push the bar of the exit door, following up with a good shove. They would then return, creeping along, bent double and with a puzzled expression on their faces—perhaps to watch someone else going through the same antics."

Of course, in these modern, professional times, percussionists are paragons of virtue who would never dream of nipping round the corner for a quick one in the middle of a symphony.

Once everyone is on stage and playing, the conductor will spend his time nodding, winking and glaring at the musicians as well as wafting that stick around. Some of these signals are simply to remind someone that, after not playing for seventeen minutes, she has to get ready to play that big fanfare. Other signals vary in meaning from "Don't forget to play this bit very quietly" to "You're fired, you butter-fingered oaf..." As far as the stick wafting is concerned, the up-down, side-to-side action of the stick is used to indicate the beats in each bar and thus the speed of the music.* The reason why most of the orchestra aren't watching most of the time is that they are busy reading the music and can only glance up occasionally to get information from the conductor.

Improvisation

Improvisation is making music up as you go along. Various sorts of music involve different levels of improvising: if you listen to a symphony written between 1800 and 1900 the total level of improvisation involved is—zero. At the other end of the scale, you get jazz musicians like Keith Jarrett, who can turn up at a venue and improvise the whole concert.

As far as improvisation is concerned, the first step is usually to play your own versions of well-known tunes. This often means playing the tune as it was originally written to start with, and then making up your own variations. The variations often involve strategies which a musician has learned, which can be applied to any tune. For example, imagine that you are one of those irritating, glazed-eyed hotel lobby pianists improvising on "My Way" to make it last until your next coffee break. If you want to make it sound romantic, play it slowly, with lots of pauses, and accompany the tune with chords split into arpeggios (where you play the notes of the chord one after another rather than all together). If you want

* I say this here because that's what is supposed to happen. In many cases, however, I have seen professional conductors simply making pointless dramatic gestures which indicate nothing to the musicians but look good to the audience.

it to sound hymn-like and spiritual, simplify the chords and play one chord for every strong beat of the rhythm. Hymns sound like this because they are specifically designed to be performed by non-expert musicians. If you want to make it sound heroic or dramatic, then don't let the music rest on the long notes; replace them with repeated notes. Of course, if you carry on doing this for too long you are likely to find a certain author creeping up behind you with a garrote in one hand and body bag in the other.

This type of variation/improvisation can, of course, be done by groups of musicians as well as solo players. Good musicians will not only arrange the notes in different ways, they will introduce new notes and change the tune. When they are improvising, the best jazz musicians merely hint at the original melody every now and then—it's like getting a glimpse of a familiar landmark when you thought you were lost. It might sound a little chaotic at times, but improvising musicians learn lots of techniques for re-establishing order whenever they want to—and can steer the audience through an emotional cycle of familiarization, disorientation, expectation and gratification. Or, in my girlfriend's view of jazz, disorientation, irritation, horror that it might never end, and relief when it does (her words, dictated to me as I write this).

Another type of improvisation involves a soloist or members of a band making up new melodies over a well-known sequence of chords or a bass line. This is epitomized in the lead guitar solo. Lead guitarists love lead guitar solos—everyone else repeats something pretty mundane, and we get to prance and pose. We've got between two and twenty-two minutes to get back to the tune and it doesn't really matter what we do in the meantime as long as it's loud and has lots of notes in it. There are examples of musically meaningful guitar solos, but the guitarists involved are just being unnecessarily clever and letting the side down. One of the commonest platforms for a lead guitar solo is the *twelve-bar blues*. The twelve-bar blues is not an allusion to the previous eleven pubs you've been to that evening, trying to forget the fact that your girlfriend just left you because you insisted on listening to too much jazz—it's a very simple musical structure which is the basis of most blues and a lot of other pop music.

The twelve bars are just twelve short time periods. As we saw in the previous chapter, music is divided up into *bars* of time and each of these time periods has several notes in it. In the context of an average blues song, a bar lasts about three seconds and contains four beats, with a slight emphasis on the first one:

dum dum *dum* dum, **dum** dum *dum* dum.

During a twelve-bar blues, the rhythm guitarist churns out a standard sequence of chords. In the simplest versions there are only three chords involved—let's call them chords X, Y and Z. Although there are quite a few varieties, in a standard twelve-bar blues you might expect the rhythm guitar player to strum chord X for four bars, then chord Y for two, back to chord X for two, Z for two and back to X for the final two of the twelve. This routine then repeats itself, and all the while the bass guitar is also supplying notes in an X–Y–X–Z–X cycle. This simple structure explains why blues bands can play after any amount of alcohol intake, and why their natural habitat is the pub. Over this straightforward musical background the lead guitarists can lark around however they want to, as long as they avoid certain notes which clash with these chords—and thus was the endless blues guitar solo born.

I have no intention of belittling blues bands, because some of them are tough-looking buggers who I wouldn't want to meet in a dark alley. So I would like to point out that, like many simple musical systems, there are lots of added complications and nuances which the best players have to master before anyone will pay them to perform.

Improvising in Western music is not a new thing; it just became unfashionable in nineteenth-century classical music. Back in the eighteenth century it was very popular: for example, one of Bach's most well-known pieces, the Brandenburg Concerto No. 3, consists of three movements but only two of them are written out in full. The written music for the middle movement is just two chords long—which would only last about ten seconds. It is possible, of course, that Bach was called away in the middle of his writing to take part in the birth or, indeed, conception, of one of his twenty children, and forgot to finish it. The more traditional historical view is that the second movement would consist of a three-minute viola solo by Bach who would then nod at the rest of the band—and they would finish off with these two chords. Nowadays, in most recordings of this piece, the musicians chicken out on the improvisation and just play the two chords—and who can blame them? I certainly wouldn't want any of my own improvised drivel sandwiched between two pieces by Bach.

Improvisation is common to all musical societies. For example, Indian traditional music concentrates heavily upon it. The training of a Western classical musician involves lots of repetition in an attempt to play the notes written by a composer correctly. Traditional Indian musical training is all about how to compose your own music on your instrument as you go along. The idea is that you have a group of notes as your basic building blocks, and you use them to improvise a piece lasting several minutes. Each group of notes, or "raga," is associated with a mood and a time of day.

The ability to improvise well is a highly respected talent and it can lead to some interesting interplay between the musicians involved. It can even become competitive, as the musicians drive each other to new heights. Speaking of competition, there are international improvisation competitions between classically trained organists. You can't cheat by playing something you composed earlier, because they hand you a brand-new tune to base your improvisation on only an hour before you begin. After you are given the tune, you have to make up a piece of music on a big church organ, in front of a crowd of other competitors and their friends. So, no pressure there then...

Whatever level you reach, improvisation is fun for the player. Even if you are a complete beginner, you can make up your own tunes really easily if you have access to a piano or similar keyboard. Use one finger from each hand and play only the black notes. This automatically gives you a pentatonic scale—and it's really quite difficult to make a horrible noise using this type of scale. If you put your foot on the right-hand pedal the notes will merge into each other, which gives a great "full" effect, but you need to raise your foot and put it down again quickly every five or six notes. This pedal lets the notes ring out for a longer time, and the notes of your melody will overlap and harmonize with each other. Every time you raise your foot you kill that group of notes. You need to do this because too many overlapping notes sound messy. The pedal on the left gives the notes a very short lifetime and is for real pianists—not for the likes of you and me.

12. Listening to Music

Concert hall acoustics

Let's imagine that you and a violin player have gone for a picnic in the countryside. If you both stop in the middle of a large, flat field and the violinist starts playing, you will find that the violin sounds a lot quieter than it does when she plays it in her living room at home.

In a room, the violin is louder because you are receiving the pressure ripples from each note several times over. You get the ripples which travel directly to your ears, plus the ones which were traveling away from you but have been turned back in your direction after bouncing off the walls, floor and ceiling. Apart from the increase in volume, this bouncing around also means that the sound is coming at you from all directions, so you feel bathed in music.

Out in the field, the violin sounds quieter because you are only receiving a double dose of the noise, once directly from the instrument, and one reflection off the ground. All the other sound travels in other directions, away from your ears.

So, the benefit of reflected sound is that we hear the music louder, and we also get the impression that we are surrounded by the music.

The size of the room and the materials that the walls, floor and ceiling are covered with determine the acoustic "*liveliness*" of the room you are in. You can check the "liveliness" of a room by clapping your hands and listening to how quickly or slowly the sound dies away. In a small room full of furniture and heavy curtains, the sound dies away almost immediately and the room is said to be acoustically "*dead*." In a larger room with hard walls, the sound bounces back and forth off the walls several times before the sound dies away, and the room is described as "lively." In a concert hall, it might take as long as a couple of seconds for this *reverberation* of your hand clap to die away. Musicians sound better in a "live" room than a "dead" one, which is why your singing sounds better when you are in the shower, surrounded by hard, tiled surfaces.

Although we enjoy the effect of sounds lasting longer because they are bouncing around the room, we want all the bounced sound from each note to overlap so that it reaches our ears as a single, long note. If the walls of the room are a long way away, the time between bounces will be too long and we won't hear a single, extended note. We will hear the note, then a gap, then the note again—the dreaded echo. Concert hall designers live in the hope that their designs will give the audience lots of pleasant reverberation, but no echoes. This is a tricky balance, because both effects are caused by sound waves bouncing off the walls, ceiling and floor.

The difference between a concert hall and a living room is that, unless you are the Queen, the concert hall will be a lot bigger, and one of the more inescapable symptoms of a larger space is that the walls are farther away from each other. In a large room like this the reflected sound has to travel a long way, from the violin all the way over to the wall and back to your eardrum. The sound which comes directly from the instrument to your ear doesn't take this detour, and therefore arrives first. So the two sounds are heard as separate events—one is an echo of the other.

In a small room, the reflected sound ripples and the direct sound all arrive at our ears at about the same time, because the round-trip to the wall and back is not much farther than the direct route.

Although the direct sound arrives first, the reflected sound is hot on its heels. If the gap between their arrival is less than forty thousandths of a second, our hearing system just assumes it's all part of the same sound. We only get a time difference above forty thousandths of a second if the round-trip taken by the reflected sound is more than 40 feet longer than the direct route from the violin to your eardrum. This could only happen in a very large space like a concert hall.

The acoustic engineers who design concert halls can reduce echo problems by putting absorbent material (which does not reflect sound very well) in places where the round-trip of the reflected sound would be a long one. They can also angle the walls so that the sound bounces around the room in the optimum way to give a "full of sound" feel without echoes.

Some concert halls have moveable or adjustable absorbing panels which can be arranged to suit different situations. For instance, you want less reflection for a public lecture or a comedian than you do for a concert. Panels like this can also be used to improve the acoustics of buildings with echo problems. One famous example of this is the ceiling of the Albert Hall in London. In this case, the curved, high ceiling used to reflect sound back down as an echo, until big, sound-absorbing "mushrooms" were installed.

Hi-fi, lo-fi or sci-fi? Home sound systems and recorded music

Microphones and speakers

Microphones and speakers are very similar devices—in fact, it wouldn't take much effort to turn any microphone into a speaker or vice versa.

The illustration opposite shows us that a microphone consists of only two important bits:

1. A small paper or plastic cone which is light enough to tremble backward and forward when sound waves hit it.
2. A device for turning this backward-and-forward trembling into an electrical signal. The electrical signal goes up and down in exactly the same way as the paper cone moves backward and forward—as you can see in the illustration. It is clear therefore that if the cone is trembling because of the wave patterns of the music, then the electrical signal is also "trembling" in the same way—we have made an electrical "copy" of the music sound waves.

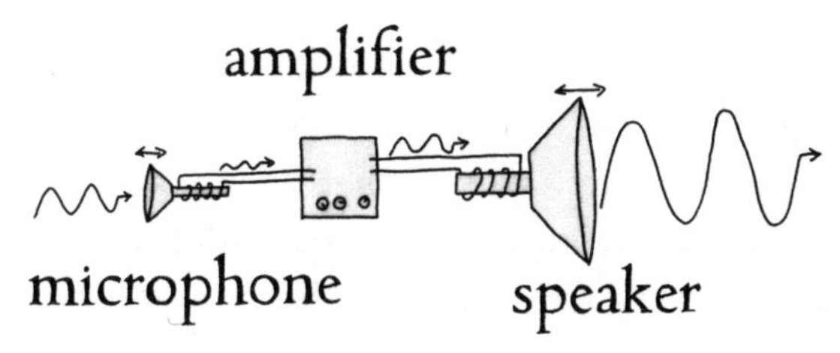

A microphone turns the backward-and-forward movement of a paper cone into an electrical signal. We can increase the power of this electrical signal by putting it through an amplifier—and use it to make the big paper cone in a loudspeaker move backward and forward—in order to reproduce the original music at a higher volume.

A speaker is just a microphone which is back to front—it has a device for turning an electrical signal into backward-and-forward trembling which is attached to a paper or plastic cone (which is usually bigger than the one in the microphone).

If we want to make a singer's voice louder, we ask them to sing into a microphone. The electrical copy of the sound is then made and passed through an amplifier which makes it much more powerful (or amplified). We then use this powerful electrical copy to make the (generally larger) paper cone in the loudspeaker tremble backward and forward in the same way as the cone in the microphone did, and the music is reproduced at a louder volume—easy-peasy.

The recording and playback of music

Instead of using a microphone and amplifier to produce louder music immediately (as at a live concert), we can take the electrical signal and store it somewhere. The electrical wiggles can, for example, be used to drive a machine to cut a wiggly groove in a plastic or metal disk. Later on we can use a machine similar to the one that cut the groove to turn the mechanical wiggling back into an electrical signal—which we can put through an amplifier to push a speaker cone backward and forward so we can hear the music again (this is exactly how vinyl records work).

There are, of course, many other ways of storing the musical information produced by the microphone, including magnetic tape, which stores wiggles of magnetization, and silicon chip data storage or compact discs (CDs), where the wiggles are converted into a stream of digital data. Every technique uses the same principle: you take the sound and convert it into stored information and then, later, you decode the information to get the sound back.

Are vinyl records better than CDs?

Ever since CDs became freely available in the 1980s a fierce debate has raged about whether or not they provide a better copy of the music than the vinyl records they replaced. Much of this debate has centered on the difference between analog and digital technology—so I would like to describe what that difference is before I go any further.

THE ANALOG/DIGITAL DIFFERENCE

To keep this part of the discussion simple, I won't talk directly about music. Instead, I will discuss the copying and reproduction of a visual image — so I can draw an example for you.

Let's say we want to copy a wavy line by analog methods and by digital techniques.

Analog reproduction

An analog recording system simply takes a wiggly line and tries to make a direct copy of it by following its curves. The principle is similar to how a cyclist follows the line down the center of a winding country lane. The accuracy of the cyclist will be determined by how fast he is going, how sharp the curves are, and how long he spent in the pub at lunchtime.

A typical example of an analog recording would be for you to copy an image using tracing paper and a pencil. It's quite easy to see how the accuracy of your trace would be improved by using exactly the right width of line and being extremely careful about it. There might be situations, however, where the line you are trying to copy wiggles to and fro too rapidly for you to follow it accurately.

Digital reproduction

Digital reproduction takes a completely different approach. The word "digital" means that a computer must reduce the task down to a series of "yes" or "no" statements. In this case the computer will divide up the page with the wiggly line on it into a lot of little squares. The computer will then point a camera at the image and ask itself "Is there a dark line in this square?" and do this for all the small squares, one at a time. The computer then stores all the "yes" and "no" answers. When the computer is asked to reproduce the image, it then prints a black square for every "yes" and a blank square for every "no." The advantage of this system is that computers can memorize zillions of "yes" or "no" answers with incredible accuracy. The information can be stored and reproduced faultlessly at any time and there is no dependence on the accuracy of moving machinery. The disadvantage of this method is that curved lines are made up of little squares—and if you don't make your squares small enough in the first place your reproduced image will not look like your original smooth wiggly line. I have demonstrated this by showing the difference between a good digital copy and one in which the small squares were too big.

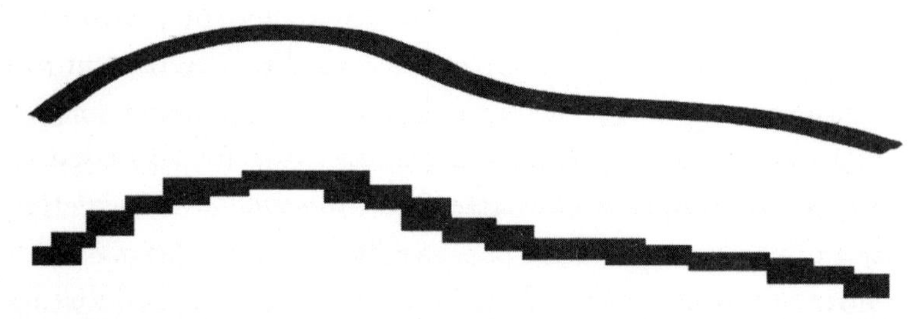

The principle of digital reproduction. Both of these images were produced digitally—using a computer to divide up the curve into a collection of black squares. If we use millions of tiny squares (as we have done in the upper image) we see a smooth curve. If the squares are too large, as they are in the lower image, we lose a lot of picture quality and the shape of the curve is only recorded approximately. Modern hi-fi equipment uses "squares" which are so tiny that we cannot hear the effect of digitization.

Now that we know the difference between analog technology (which is used to produce vinyl records) and digital technology (CDs), we can answer our original question: "Are vinyl records better than CDs?" And the answer is... very *very* few people can tell the difference between the two (as long as the vinyl records are in perfect condition—and we use good equipment in both cases). This point was proved by a couple of music psychologists (Klaus-Ernst Behne and Johannes Barkowsky) in 1993. They took 160 people who were seriously into music systems and who had strong opinions about the CD/vinyl debate and made them listen to both types of music reproduction. Only four out of the 160 could actually identify whether or not they were listening to a CD—even though the vinyl fans all began the test thinking that CDs sounded "shrill and dead" compared to the "warm" sound of vinyl. Also, don't forget that these weren't just average listeners—they were keen, opinionated enthusiasts. The number of average listeners who could tell the difference between the sound of a CD and that of a vinyl record was probably less than one in a hundred—and that's back in 1993. Improvements in technology since then have undoubtedly reduced this number and rendered the comparison irrelevant.

Much of the CD/vinyl debate can probably be attributed to technology nostalgia, which dates back to cave-dwellers having heated arguments about the superiority of bronze arrow heads compared with the newfangled iron ones. Back in the 1930s music fans were complaining that, because the new recording techniques could handle loud and quiet music, they missed the excitement of the distortion which took place in orchestral climaxes on their older records. Later, in 1963, a review of the latest technology (RCA Dynagroove recording) noted that some listeners found the new, smoother sound too sterile. Personally I think that the difference between vinyl and CD sound is bound to be irrelevant compared to variables like the ticking of the central heating, traffic noise, and the plaintive voice in the background asking if this jazz is going to be playing for much longer....

The difference between CDs and MP3 technology

Imagine we are at a concert watching our favorite band (the Psychedelic Death Weasels) playing their epic rock ballad, "Is my cocoa ready yet, luv?"

During the quiet, romantic verses we can clearly hear all the instruments, including the acoustic guitar which is being played by the singer. However, when the band plays the heavy rock chorus, all we can hear are the bass, drums and electric guitar. We can see that the singer is still playing his acoustic guitar but the sound he is making is completely drowned out by the other, louder instruments.

If this track was being recorded onto CD, every sound made by every instrument would be faithfully recorded as digital information—even the inaudible music produced by the acoustic guitar during the hard rock chorus. As far as the digital recording process is concerned, the same amount of data will be collected for the "hidden" guitar as for the much louder instruments. You wouldn't hear these "hidden" sounds at the concert or on the CD, so the faithful collection of this data is pointless—but the recording equipment does it automatically because it doesn't know how to pick and choose.

This "drowning out" or hiding of one instrument by another happens all the time during the performance of any type of music. Sometimes (as in the example above) one instrument is hidden for several seconds or even minutes. In many cases, however, instruments are only hidden for a fraction of a second—for example, a loud drum note might drown out a whole band or orchestra.

Apart from these hidden sounds, a CD also contains a lot of information that we simply cannot hear—frequencies which are too high or too low for the human ear. As we saw in earlier chapters, musical notes are made up of a family of related frequencies: the fundamental frequency, twice that frequency, three times that frequency, four times, five times, etc. If we play the highest notes on certain instruments, some of their harmonics will be out of our hearing range. Similarly, some combinations of low notes produce subsonic waves which are too low for human ears (although you can sometimes feel them). On a CD these inaudible parts of the notes are all stored and played back—even though we can't hear them.

In the 1980s and '90s a bunch of ridiculously intelligent scientists and engineers developed a method of using computers to identify all the hidden and inaudible information on music CDs. Once it was identified, it could be discarded and the music could be re-recorded without all that redundant information. Approximately 90 percent of the information on a CD can be discarded in this way to produce an MP3 file. This, of course, means that you could record ten CDs' worth of music on one CD. Alternatively you can store and play back the music as digital information on a computer or personal stereo (iPods, etc.). Although MP3 technology discards most of the information that came out of the original musical performance, the average listener cannot tell the difference between a CD and an MP3 playback.

Home music systems

Music system enthusiasts, or "audiophiles," can spend over a year's salary on their systems—and if that's what they want to do it's fine by me. On the other hand, you can buy a music system which will approximately match the performance of your ears for about $1,000. I would advise you to go to a specialist music system shop which specifically advertises their goods as being high quality but cheap. I also recommend buying stuff second-hand from an enthusiast (enthusiasts tend to upgrade every couple of years and the equipment they buy and sell is always high quality). Up to about the $1,000 level, new equipment generally increases in quality as the price goes up—but it's best to have a hi-fi enthusiast giving you advice or refer to hi-fi magazines for their "best under $1,000" choices. Between $1,000 and $3,000 the increases in sound quality can be difficult to spot and above $3,000 the money/sound-quality correlation disappears altogether, as far as I can tell. The different systems might sound different—but it's very difficult to identify whether one is actually better than another. (It is possible to spend over $1,500 per yard for audio cables and I would be very interested to meet anyone who could tell the musical difference between such cables and ones which cost only a few dollars per yard.)

Having bought your equipment you now have two choices:

1. You can employ an acoustic technician and an architect. For $75,000 they will build you a special listening room and, when they have finished, they will make the biggest difference to the sound by trying the speakers in different places and moving the furniture about; or
2. You can save yourself $75,000 by taking the equipment home to a normal room, trying the speakers in different places and moving the furniture about.

The main difference between "pop" and "serious" music

Most melodies are only a few seconds long and what you do with this tiny bit of material is the source of the main difference between so-called pop and serious music. The following comments are not intended to make one genre sound better than the other—I love them both. Also, I am going to make some disgracefully broad generalizations in order to make my point.

A pop music composer will take two or three short musical ideas and make them into a three-minute song by playing them one after another. They will, for example, take tune "A" and make it into the chorus of a song, and tune "B" will be made into the verse. The song will then take the form: introduction—verse—chorus—verse—chorus—guitar solo—verse—chorus—finale.

This technique of continuously repeating the two tunes has several effects on the listener (assuming that the tunes are good ones):

1. It's easy to remember the tunes.
2. It's easy to get rapidly addicted to the tunes.
3. It's easy to get bored with the whole thing after it's been played thirty or forty times.

One common aspect of pop or rock songs is their use of a *hook*—a short, repeated musical phrase which is easily remembered. These can be melodic or rhythmic and usually last between seven and twelve seconds. Sometimes the song begins with the hook, as in the first five notes of "Whole Lotta Love" by Led Zeppelin or the first line of "Baby Love" by The Supremes. In other cases you have to wait a while before you hear the hook, which is what happens in "Momma Told Me Not to Come" by Three Dog Night, or "Teenage Dirtbag" by Wheatus. In all these cases the title words form part of the hook.

Composers of "serious" music are not averse to hooks either—just look at Beethoven's Fifth Symphony, with its "Da Da Da Daah" opening. On the whole, though, "serious" music composers are a little more cautious and miserly with their material than pop composers. Their aim is to take two or three tunes and use them as the basis for a piece of music which might last between ten and a hundred minutes. This is done by using techniques such as breaking the tunes up and playing with the fragments; merging one tune into the other; playing one tune as the accompaniment to the other; and hinting that the tune is on its way. A composer of a long piece of music might use fragments of the main tune as landmarks to build up expectation—and expectation is much more important in long pieces than it is in short pop songs.

Many people, particularly classical music professionals, think that the listener retains some sort of appreciation of the key in which the piece begins and can sense the "homecoming" when we return to that key, as the music often does toward the end of a piece of classical music. I think this is a bit far-fetched. It's expecting too much of the listener's memory unless she has perfect pitch. Some studies have indicated that our memory of what's going on harmonically only stretches back a minute or so, and this seems much more realistic. People *will* remember melodies or effects such as drum motifs from earlier in the music and will be pleased if they hear fragments of things they recognize. As far as the harmony goes, however, my analogy about different keys being like the rungs of a hamster wheel comes back into play. How are we supposed to know if the final rung is the rung we started on?

I think the experience of listening to a long piece of music can be likened to walking on a decorative carpet which is being rolled out in front of you and is being rolled back up a couple of yards behind you. As you walk forward new patterns and images appear and some of these will stick in your mind. Let's say you saw an image of a tiger a few minutes ago and you can see the beginnings of a tiger tail coming up. The composer could choose to satisfy your expectation by showing you another view of the tiger. Or he could decide to surprise you—that "tiger tail" might actually be the rear end of a snake. If you listen to a piece several times and begin to know it better, you get fewer surprises and a better view of the whole carpet. You can also take a lot of pleasure from recognizing various landmarks as you go.

A skillful composer's job is to create expectations and then either to satisfy or frustrate them. But the composer cannot and must not try for continuous excitement. As in any storytelling, or even a fireworks display, you deliberately add some calmer passages, so that the important moments make a greater impact.

These techniques for longer pieces result in music which is less easy to love at first hearing, but which seems to improve each time you hear it.

And finally . . .

Whether you like pop music, heavy metal or classical, you might find it interesting to listen to your favorite pieces and follow just one instrument at a time. Try playing your favorite pop song a few times while you listen only to what the bass guitar is doing; then do the same for the other instruments. You can learn a lot about how a piece is organized in this way and you will be really *listening* to the music rather than just hearing it.

I would like to finish with the best bit of advice that one listener can give to another. Whatever your present tastes, there are probably several other types of music out there which would bring you a lot of pleasure if you became more familiar with them. My advice is this: try adding a bit of randomness to your listening pattern, and give each new type of music a fair trial. If you are into heavy metal, try some folk music; if you love Mozart, try Dolly Parton. Musical genres are not exclusive, and one easy way to increase the amount of fun in your life is to expand the range of your listening.

Fiddly Details

A. Naming and identifying intervals

At various points in the book I have mentioned that the jump in pitch between any two notes is called an interval. The interval we have discussed most is the octave—the jump in pitch that corresponds to a doubling of the frequency of vibration of the note. All the other intervals also have names, some of which I have referred to already. The table opposite presents a list of interval names along with the size of the interval in semitones. You can also see three photos of a pianist playing two notes a *fourth* apart. In each case we start on any lower note and count up to the note which is five semitone steps higher (counting the black notes as well as the white ones). I have included three illustrations here in order to make it clear that it doesn't matter which note you start on—the one five semitone steps up will always be a fourth higher, and similarly the one nine semitones up will always be a *major sixth* higher.

After years of musical training you get to recognize each of these intervals and you can write down any tune which comes into your head (start on any note, up a fifth, down a major third, etc.). But there is a way of identifying musical intervals which anyone can manage. All you have to do is learn the names of the intervals which occur at the beginning of several songs. I have made a list of suitable songs in the table below. The first two notes of the songs identified in this list give you the interval in its rising form (unless

I have stated otherwise). Most tunes start with a rising interval, so examples are plentiful and you might want to substitute other songs for your own list. You might also want to collect twelve descending intervals from other songs.

So now, when you're bored, sitting at an airport, you can identify the interval of whichever annoying "bing bong" noise they are using by matching it to one of these songs.

Size of interval	**Name of interval**	**Song to identify rising interval**
1 semitone	minor second or semitone	"I left my heart in San Francisco"
2 semitones	major second or tone	"Frère Jacques" or "Silent night"
3 semitones	minor third	"Greensleeves" or "Smoke on the Water" (by Deep Purple)—first two guitar notes
4 semitones	major third	"While shepherds watched their flocks by night" or "Kum ba yah"
5 semitones	fourth	"Here comes the bride" or "We wish you a merry Christmas"
6 semitones	diminished fifth (or augmented fourth)	"Maria" from *West Side Story*
7 semitones	fifth (or perfect fifth)	"Twinkle twinkle little star..." (the jump between the two twinkles)
8 semitones	minor sixth (or augmented fifth)	The theme from *Love Story* begins with this interval falling then rising
9 semitones	major sixth	"My bonnie lies over the ocean..."
10 semitones	minor seventh	"Somewhere" from *West Side Story*
11 semitones	major seventh	"Take on me" (song by the band a-ha)

Size of interval	Name of interval	Song to identify rising interval
12 semitones	octave	"Somewhere over the rainbow" from *The Wizard of Oz*
13 semitones	minor ninth	
14 semitones	major ninth	
15 semitones	minor tenth	
16 semitones	major tenth	
17 semitones	eleventh (or an octave and a fourth)	

Three photos of a pianist playing two notes a fourth apart. It doesn't matter which note you start on—the two notes are always separated by a gap of five semitones.

B. Using the decibel system

1. The decibel system is a method of comparing the difference in volume between two sounds. These differences are measured in the following way:

- If the difference between two sounds is 10 decibels then one sound is twice as loud as the other.

- If the difference is 20 decibels then one sound is 4 times as loud as the other.
- If the difference is 30 decibels then one sound is 8 times as loud as the other.
- If the difference is 40 decibels then one sound is 16 times as loud as the other.
- If the difference is 50 decibels then one sound is 32 times as loud as the other.
- If the difference is 60 decibels then one sound is 64 times as loud as the other.
- If the difference is 70 decibels then one sound is 128 times as loud as the other.
- If the difference is 80 decibels then one sound is 256 times as loud as the other.
- If the difference is 90 decibels then one sound is 512 times as loud as the other.
- If the difference is 100 decibels then one sound is 1,024 times as loud as the other.
- If the difference is 110 decibels then one sound is 2,048 times as loud as the other.
- If the difference is 120 decibels then one sound is 4,096 times as loud as the other.

2. When using rule 1 (above) it doesn't matter which decibel number you start on. For example, the difference between 10 decibels and 20 decibels is 10 decibels so 20 decibels is twice as loud as 10 decibels. But the difference between 83 decibels and 93 decibels is also 10 decibels—so 93 decibels is twice as loud as 83 decibels. Similarly, 72 decibels is 16 times as loud as 32 decibels (because the difference between 72 and 32 is 40 decibels).

3. Although the decibel system should only be used to compare the relative loudness of two noises (as in rules 1 and 2 above), many people use decibels as a definite measure of the loudness of a single

noise. In this case they are not saying "a large bee creates a noise of 20 decibels"; what they are actually saying is "a large bee creates a noise which is 20 decibels louder than the quietest noise we can hear." The quietest noise we can hear is called the "threshold of hearing," so we are really saying "a large bee creates a noise which is 20 decibels louder than the threshold of hearing." When people appear to be using decibels in a non-comparative way (e.g., "The loudness of that motorbike is 90dB") it's just because they haven't bothered to include the phrase "louder than the threshold of hearing"—it's taken for granted.

C. Tuning an instrument to a pentatonic scale

If you are doing this as an experiment it would be convenient to use a guitar, as it's the commonest six-string instrument. On a guitar, the strings are traditionally numbered 1 to 6 with 6 being the thickest one and the lowest note. Unfortunately, this is the opposite of the numbering system I used in chapter 8. I tried reversing the numbers in the chapter to match the guitar system, but it spoiled the clarity of the chapter. So I'm going to give you instructions twice—once following the numbering system in chapter 8, and once using the traditional guitar numbering for the strings. You can use either set of instructions—they both give exactly the same result.

If you tune a guitar to a pentatonic scale, you will change the difference between the thickest string and the thinnest from two octaves to one octave. If you make the thickest string give its usual note before you start, this will mean that the thinner strings will be very slack by the time you have finished. This doesn't matter much if you are doing this out of curiosity and you are going to return the guitar to its normal tuning in a few minutes. If, on the other hand, you are doing this as a long-term plan or as a demonstration for students, I'd recommend that you tighten the thickest string to give a higher note before you start and/or change the thinner strings for thicker ones.

Re-tuning like this takes a while if you start with a normally tuned guitar, because the strings resent being de-tuned so much and take a long time to settle down to their new tighter or slacker tensions. I also suspect that the unbalanced tension of the strings will cause the neck to bend if you leave a guitar tuned like this for several days.

So, here we go. One finger of one hand will be used to pluck the strings individually, and one finger of the other hand (called the "non-plucking finger," below) should gently rest on the string—as shown in the photo in chapter 8.

Each string can produce four notes easily:

- the natural note of the open string (which we will call "open")
- the "octave above" pinged note—created by pinging the string with your non-plucking finger on the middle of the string (on a guitar, the middle of the string is directly above the twelfth fret)
- the "double octave above" note—created by pinging the string with your non-plucking finger one quarter of its length from either end—directly above the fifth fret
- the "new" note—created by pinging the string with your non-plucking finger one third of its length from either end—directly above the seventh fret.

If we call the thickest string number 1

First of all we tune the thickest string (string 1) to a higher than normal note (to stop the thinner strings being too slack when we have finished).

Then the "octave above" note of string 1 should match the "open" note of string 6.

The "new" note of string 1 should match the "octave above" note of string 4.

The "new" note of string 4 should match the "double octave above" note of string 2.

The "new" note of string 2 should match the "octave above" note of string 5.

The "new" note of string 5 should match the "double octave above" note of string 3.

All done—instant Oriental sound.

If we call the thickest string number 6 (normal guitar numbering)

First of all we tune the thickest string (string 6) to a higher than normal note (to stop the thinner strings being too slack when we have finished).

Then the "octave above" note of string 6 should match the "open" note of string 1.

The "new" note of string 6 should match the "octave above" note of string 3.

The "new" note of string 3 should match the "double octave above" note of string 5.

The "new" note of string 5 should match the "octave above" note of string 2.

The "new" note of string 2 should match the "double octave above" note of string 4.

All done—instant Oriental sound.

D. Calculating equal temperament

As I said in chapter 8, Galilei and Chu Tsai-Yu found that calculating the equal temperament system is pretty easy once you have presented the problem clearly and logically:

1. A note an octave above another must have twice the frequency of the lower one. (This is the same as saying that if you use two identical strings one must be half the length of the other—the frequency of the note produced by a string goes up as the string gets shorter and half the length gives double the frequency.)
2. The octave must be divided up into twelve steps.
3. All the twelve steps must be equal. (If you take any two notes one step apart, then the frequency ratio between them must always be the same.)

Let's look at an example to make the calculation clearer. In this example we will make every string 90 percent as long as its longer neighbor—and all strings are made of the same material and are under the same amount of tension.

1. Let's make our longest string 24 inches long.
2. String number 2 is 90 percent of the length of string 1 (i.e., 21.6 inches long).
3. String number 3 is 90 percent of the length of string 2 (i.e., 19.4 inches long).
4. String number 4 is 90 percent of the length of string 3 (i.e., 17.5 inches long).
5. Continue until you reach string 13.

This example shows how the percentage shortening system works, but unfortunately we picked the wrong percentage. There is too much of a jump between strings. We want the thirteenth string to be half as long as the original string so that it will produce a note an octave above it. However, if you make every string 90 percent of the length of the previous one, your thirteenth string will be far shorter than you want it to be. So what percentage should we use for shortening our strings?

This is where Galilei and Chu Tsai-Yu come in handy. They

calculated* exactly the correct percentage to make string 13 half the length of string 1. And the answer is . . . 94.38744 percent. Or, to put it another way, you need to remove 5.61256 percent of the length of any string to find the length of its shorter neighbor.

So now let's do the calculation using this correct percentage:

1. String 1 is 24 inches long.
2. String 2 is 94.38744 percent as long as string 1—i.e., 22.65 inches.
3. String 3 is 94.38744 percent as long as string 2—i.e., 21.38 inches.
4. String 4 is 94.38744 percent as long as string 3—i.e., 20.18 inches.
5. String 5 is 94.38744 percent as long as string 4—i.e., 19.05 inches.
6. String 6 is 94.38744 percent as long as string 5—i.e., 17.98 inches.
7. String 7 is 94.38744 percent as long as string 6—i.e., 16.97 inches.
8. String 8 is 94.38744 percent as long as string 7—i.e., 16.02 inches.
9. String 9 is 94.38744 percent as long as string 8—i.e., 15.12 inches.

---

* Galilei and Chu Tsai-Yu probably started by calculating the increase in frequency between two adjacent notes. From this you can work out how much shorter the higher string should be. I have used the shortening of strings because it makes the discussion easier to follow. Our two wise men calculated that we needed an increase in frequency of 5.9463 percent between two adjacent strings: for example, if G has a frequency of 392Hz, then the note one semitone up (G sharp) has a frequency of 105.9463 percent of 392, which is 415.3Hz. To achieve this, if the strings are otherwise identical, the G sharp string will have to be 94.38744 percent the length of the G string.

10. String 10 is 94.38744 percent as long as string 9—i.e., 14.27 inches.
11. String 11 is 94.38744 percent as long as string 10—i.e., 13.47 inches.
12. String 12 is 94.38744 percent as long as string 11—i.e., 12.71 inches.
13. String 13 is 94.38744 percent as long as string 12—i.e., 12.00 inches.

Now the thirteenth string is half as long as the first string, which is what we wanted. Also, the relative length of any string compared to its neighbor is always the same—so it doesn't matter which string you start your tune on, the same sequence of up and down jumps will give you the same tune (but the whole tune will be higher or lower in pitch).

E. The notes of the major keys

In the following list, "♭" means "flat" and "#" means "sharp."

A major: A, B, C#, D, E, F#, G#
B♭ major: B♭, C, D, E♭, F, G, A
B major: B, C#, D#, E, F#, G#, A#
C major: C, D, E, F, G, A, B
D♭ major: Db, E♭, F, G♭, A♭, B♭, C
D major: D, E, F#, G, A, B, C#
E♭ major: E♭, F, G, A♭, B♭, C, D
E major: E, F#, G#, A, B, C#, D#
F major: F, G, A, B♭, C, D, E
F# major: F#, G#, A#, B, C#, D#, E# (or G♭ major: G♭, A♭, B♭, C♭, D♭, E♭, F)

G major: G, A, B, C, D, E, F#
A♭ major: A♭, B♭, C, D♭, E♭, F, G

Note: F# is the same as G♭—so we could have either key in this case. In every other case where the key could have one of two names (e.g., D♭ and C#), we generally choose the one with the least sharps or flats in the key signature. D♭ involves five flats but its alternative, C#, would have seven sharps—so we generally use D♭. There is no such clear choice in the case of F#/G♭ because F# has six sharps and G♭ has six flats.

Bibliography

Books

The physics of music

J. Backus, *The Acoustical Foundations of Music*, 2nd edn, W. W. Norton, 1977

W. Bragg, *The World of Sound*, G. Bell and Sons, 1927

M. Campbell and C. Greated, *The Musician's Guide to Acoustics*, new edn, Oxford University Press, 1998

H. Helmholtz, *On the Sensations of Tone*, 2nd edn, Dover Publications, Inc., 1954

D. M. Howard and J. Angus, *Acoustics and Psychoacoustics*, 2nd edn, Focal Press, 2000

I. Johnston, *Measured Tones — The Interplay of Physics and Music*, 2nd edn, Taylor & Francis, 2002

S. Levarie and E. Levy, *Tone — A Study in Musical Acoustics*, 2nd edn, Kent State University Press, 1980

J. R. Pierce, *The Science of Musical Sound*, 2nd revised edn, W. H. Freeman and Co., 1992

J. G. Roederer, *The Physics and Psychophysics of Music — An Introduction*, 3rd edn, Springer-Verlag, 1995

C. Taylor, *Exploring Music — The Science and Technology of Tones and Tunes*, Taylor & Francis, 1992

The psychology of music

J. Dowling and D. L. Harwood, *Music Cognition*, Academic Press Inc., 1986

D. Huron, *Sweet Anticipation—Music and the Psychology of Anticipation*, MIT Press, 2006

P. N. Juslin and J. A. Sloboda (eds), *Music and Emotion—Theory and Research*, Oxford University Press, 2001

D. L. Levitin, *This Is Your Brain on Music: The Science of a Human Obsession*, Dutton Books, 2006

B. C. J. Moore, *An Introduction to the Psychology of Hearing*, 3rd edn, Academic Press, 1989

O. Sacks, *Musicophilia—Tales of Music and the Brain*, Picador, 2007

J. A. Sloboda, *The Musical Mind—The Cognitive Psychology of Music*, Clarendon Press, 1985

General

G. Abraham (ed.), *The Concise Oxford History of Music*, Oxford University Press, 1985

F. Corder, *The Orchestra and How to Write for It*, J. Curwen and Sons, 1894

R. W. Duffin, *How Equal Temperament Ruined Harmony*, W. W. Norton and Co., 2007

A. Einstein, *A Short History of Music*, 3rd edn, Dorset Press, 1986

H. Goodall, *Big Bangs: The Story of Five Discoveries that Changed Musical History*, Chatto and Windus, 2000

G. Hindley (ed.), *The Larousse Encyclopedia of Music*, Hamlyn, 1971

B. McElheran, *Conducting Technique: For Beginners and Professionals*, 3rd edn, Oxford University Press, 2004

R. and J. Massey, *The Music of India*, Kahn and Averill, 1976

W. Piston, *Orchestration*, Victor Gollancz, 1978 (originally published, 1955)

A. Ross, *The Rest Is Noise: Listening to the Twentieth Century*, Fourth Estate, 2008

C. Sachs, *A Short History of World Music*, Dobson Books, 1956

S. Sadie (ed.), *Collins Classical Music Encyclopedia*, HarperCollins, 2000

P. A. Scholes (ed.), *The Oxford Companion to Music*, 2nd edn, Oxford University Press, 1939

R. Smith Brindle, *Musical Composition*, Oxford University Press, 1986

E. Taylor, *The AB Guide to Music Theory*, Parts 1 and 2, The Associated Board of the Royal Schools of Music, 1989/1991

H. H. Touma, *The Music of the Arabs*, Amadeus Press, 1996

H. J. Wood, *My Life of Music*, Victor Gollancz, 1938

Journal Papers

D. Deutsch and K. Dooley, "Absolute pitch among students in an American music conservatory: Association with tone language fluency," *J. Acoust. Soc. Am.* 125(4), April 2009, pp. 2398–2403

R. W. Duffin, "Just Intonation in Renaissance Theory and Practice," *Music Theory Online*, Vol. 12, No. 3, October 2006

J. Powell and N. Dibben, "Key–Mood Association: A Self-Perpetuating Myth," *Musicae Scientiae*, Vol. IX, No. 2, Fall 2005, pp. 289–312

Acknowledgments

I would, of course, like to take all the credit for the good bits of this book, and blame all the bad bits on someone else—preferably someone I don't like who lives miles away. Unfortunately, this sort of behavior is generally frowned upon, so I had better come clean and admit that I had a lot of help.

My editor, Helen Conford, was endlessly helpful in keeping me on the straight and narrow and trying to get me to produce a readable piece of work. Without the enthusiasm of my agent, Patrick Walsh, the whole thing would have fallen at the first fence. I would also like to thank Sarah Hunt-Cooke for her impressive wheeler-dealing work on the project. Also, thanks to Jane Robertson for her copy-editing and to Rebecca Lee for her editorial coordination.

Many thanks to Tracy Behar and Michael Pietsch of Little, Brown for lots of useful ideas and guidance, and to their colleagues Christina Rodriguez and Peggy Freudenthal for all their editorial and production work.

Thanks also to my friends and family who read through the shabby initial drafts and gave lots of useful feedback and ideas, especially Dr. Nikki Dibben, Utkarsha Joshi, Libby Grimshaw, Angela Melamed, Dr. Steve Dance, Tony Langtry, Mini Grey (and Herbie), Rod O'Connor, Dr. Donal McNally, Mike Smeeton, Clem Young, Dr. John Dowden, Whit, Arthur Jurgens and, of course, Queenie (my mother). Thanks are also due to my composition tutor, Dr. George Nicholson.

Special thanks to Jo Grey for the photographs.

A gold-medal-order-of-merit is due to my girlfriend Kim Jurgens for her excellent proof-reading, encouragement and help with the artwork—and for preventing me from throwing my laptop and/or printer out of the dining-room window on numerous occasions.

Finally, the platinum award for suggestions, corrections and back-up information is awarded to my unfeasibly well-informed friend John Wykes.

Having handed out these justifiable honors, I see no reason why I should go the whole hog and admit responsibility for any errors, omissions or other bad stuff. I think any blame involved should be borne by my secondary school geography teacher, Mr. Nigel Jones, of 14b Montebank Close, Eccles, Lancashire. Please send any negative comments, lawsuits or demands for compensation directly to him at the above address.

Index

Page numbers in **bold** indicate a main reference, *italics* indicate an illustration

About the Author

Dr. John Powell is a physicist and a classically trained musician, with naturally curly hair. He has given lectures at international laser conferences and played guitar in pubs in return for free beer. He prefers the latter activity. Dr. Powell decided to write this engaging and accessible introduction to the science and psychology behind music when he discovered that all the existing books on the subject gave him a headache. He holds a master's degree in music composition and a PhD in physics, and has taught physics at the Universities of Nottingham and Lulea (Sweden) and musical acoustics at Sheffield University. He lives in Nottingham, England.